Sustainable Development Goals Series

The **Sustainable Development Goals Series** is Springer Nature's inaugural cross-imprint book series that addresses and supports the United Nations' seventeen Sustainable Development Goals. The series fosters comprehensive research focused on these global targets and endeavours to address some of society's greatest grand challenges. The SDGs are inherently multidisciplinary, and they bring people working across different fields together and working towards a common goal. In this spirit, the Sustainable Development Goals series is the first at Springer Nature to publish books under both the Springer and Palgrave Macmillan imprints, bringing the strengths of our imprints together.

The Sustainable Development Goals Series is organized into eighteen subseries: one subseries based around each of the seventeen respective Sustainable Development Goals, and an eighteenth subseries, "Connecting the Goals," which serves as a home for volumes addressing multiple goals or studying the SDGs as a whole. Each subseries is guided by an expert Subseries Advisor with years or decades of experience studying and addressing core components of their respective Goal.

The SDG Series has a remit as broad as the SDGs themselves, and contributions are welcome from scientists, academics, policymakers, and researchers working in fields related to any of the seventeen goals. If you are interested in contributing a monograph or curated volume to the series, please contact the Publishers: Zachary Romano [Springer; zachary.romano@springer.com] and Rachael Ballard [Palgrave Macmillan; rachael.ballard@palgrave.com].

Rashed Al Mahmud Titumir
Editor

Sundarbans and its Ecosystem Services

Traditional Knowledge, Customary Sustainable Use and Community Based Innovation

palgrave
macmillan

Editor
Rashed Al Mahmud Titumir
Department of Development Studies
University of Dhaka
Dhaka, Bangladesh

Unnayan Onneshan
Dhaka, Bangladesh

ISSN 2523-3084 ISSN 2523-3092 (electronic)
Sustainable Development Goals Series
ISBN 978-981-19-2999-1 ISBN 978-981-19-3000-3 (eBook)
https://doi.org/10.1007/978-981-19-3000-3

Cover illustration: ALAMY B093NN

This Palgrave Macmillan imprint is published by the registered company Springer Nature Singapore Pte Ltd.
The registered company address is: 152 Beach Road, #21-01/04 Gateway East, Singapore 189721, Singapore

Dedicated to
The members of three cooperatives of traditional resource users of Sundarbans— without them this work would not have been done!

Koyra Bonojibi Bohumukhi Unnayan Samity
Harinagar Bonojibi Bohumukhi Unnayan Samity
Munda Adivasi Bonojibi Bohumukhi Unnayan Samity

Preface

The Sundarbans, the largest single-tract contiguous mangrove forests in the world, is a blessing for the coastal belt of Bangladesh. The home of Royal Bengal Tiger works as a shield against natural calamities and nurtures lives and livelihood in the region. Since antiquity, these dense forests have provided the people living along the coast with infinite compassion as a result of which the Indigenous People and Local Communities (IPLCs) find an intertwined relationship with this bountiful and the largest mangrove.

The forest shares an intense, interdependent relationship with the indigenous and local communities living around it. People who depend on the Sundarbans for their livelihood, also contribute to the conservation of the Sundarbans' biodiversity and heritage. Nevertheless, due to various man-made reasons, the forest is being pushed on the verge of destruction. As a result, the future of the forest and the people dependent on it is endangered.

The Sundarbans is gradually degrading due to the massive extraction of its resources. The inappropriate expropriation has triggered rapid destruction and has failed to take into consideration, sustainability and conservation of the forest resources. This urbane mindset of resource exploitation stands in stark contrast to the Indigenous Local Knowledge (ILK) system which stems from age-old practice of conservation-prone and sustainable use of forest resources.

This book attempts to explore the intricacies of the largest surviving mangrove forest in the world by studying various aspects related to the forest; some of these aspects are the traditional and customary knowledge and rules and beliefs practised by the Traditional Resource Users (TRUs) residing around the Sundarbans. Knowledge of these practices are essential for conserving the natural resources of the forest that have been ascertained over a period of time through keen observation, proving the efficacy of such practices towards sustainable use of natural resources. In essence, this book is a brief glossary of ILK and attempts to link the mainstream conservation practices of the forest's terrestrial, coastal and marine ecosystems to the existing Traditional Knowledge (TK).

The sanguinity of divulging such knowledge to protect the security, sanctity and serenity of the forest through sustainable use of its resources depends on exploring its current state and multiple values that require an in-depth understanding of the interdependent relationship that Sundarbans share with the people dependent on it. A similar attempt is needed to identify the factors affecting the well-being of both humans and the forest, as a habitat. Through proper investigation, the interdependence of humans and the Sundarbans has been explored that has been found essential for each other's existence and future growth. This book attempts to draw on the work-experience and data reservoir collated by the *Unnayan Onneshan*—a Dhaka-based multidisciplinary, independent research organisation working in the Sundarbans for almost a decade to bring forth a scientific narrative of this majestic forest to the researchers and general readers as a whole.

This book is an inter-disciplinary inquiry undertaken in the Sundarbans by adopting multiple evidence-based approach in order to bring participatory insights of the indigenous people and local communities, coupled with scientific knowledge and heterodox perspectives including Anthropology, Ecology, Economics and Political Economy in a shared platform. It presents a comprehensive story of the Sundarbans—a natural World Heritage Site and a Ramsar Site of international importance and introduces a new theoretical framework by combining theory, praxis and policy in a triangulation method and advocates for ensuring rights-oriented universal social entitlements, provision of livelihood security and promotion of human–nature cooperation.

All the chapters of this book have been written by adopting Multiple Evidence-Based (MEB) approach that includes both primary and secondary data and brings heterodox multidisciplinary perspectives to a common platform. The primary data has been collected through participatory approaches such as Focus Group Discussions (FGD), unstructured interview, Key Informant Interview (KII) and Participatory Rural Appraisal (PRA) tools like social mapping, impact assessment by the indigenous and local people. Public Participation Geographic Information System (PPGIS) has also been conducted for ecological and landscape valuation mapping, resource mapping, etc. In addition, few photos of Sundarbans—taken by the researchers on the ground during field work have been used in several chapters. These participatory approaches have been drawn towards knowledge, views and understanding of the TRUs. Similarly, secondary scientific evidence has been collected from numerous articles, journals, books, newspapers, etc.

Chapter 1 of the book presents a comprehensive overview of the Sundarbans. This includes its location and a brief explanation of its flora and fauna. It goes on to describe the biodiversity and resource system in the Sundarbans. In addition, the chapter attempts to develop an alternative conceptualisation of the multiple values and the nature of the TRUs, emphasising this age-old value system. During the course of the chapter, it is established how the Sundarbans is considered to be the TRUs' mind. Later in the chapter, the approach of the current management system and associated problems have been discussed. Finally, the introductory chapter ends with a methodology of the study and a brief outline of the upcoming chapters in the book.

Chapter 2 opens with exploring the trends in biodiversity and resources. It examines current terrestrial ecosystems and later delves into the vulnerability profile of several valuable resources. The chapter suggests how this forest is classified as a unique Socio-Ecological Production Landscape and Seascape (SEPLS) comprising terrestrial forestry, marine and coastal ecosystem that offers myriad livelihood opportunities. It goes on to show the degrading state of the terrestrial ecosystem of the forest through a DPSIR (Driver, Pressure, State, Impact and Response) framework. Lastly, the chapter investigates the management approaches along with associated problems and challenges being faced that results in resource degradation. It demonstrates that biodiversity of terrestrial ecosystem of this mangrove forest is tremendously declining and there has been massive degradation of forest coverage and its density in last two decades.

Chapter 3 focuses on the state of biodiversity in marine and coastal ecosystems of the Sundarbans. While doing so, it takes into account the perspectives and opinions of the TRUs. The chapter demonstrates how aquatic resources are degrading gradually and also depicts the changing patterns of value in marine and coastal ecosystem. A DPSIR framework is also developed in this chapter to depict the key causes of damage to the aquatic resources and its impact. The chapter reveals that aquatic resources are declining mainly due to the exaggerated extraction of resources by unsustainable means accelerated by profit making impulses of people. As a result, both aquatic floral and faunal species significantly have declined in terms of quality and quantity.

Chapter 4 explains the traditional knowledge and explores the customs of the local forest people in context to the terrestrial area of the forest. It studies the age-old norms and methods followed by the TRUs that is the *bawali* (golpata collectors), *mouals* (honey/wax collectors) and crab collectors, the cultural beliefs of the traditional resource users and indigenous people that have been at the foundations of their mode of living and daily practices. The chapter goes on to study the innovations in livelihood along with agriculture and aquaculture. Finally, the chapter looks into the roles of TRUs towards conservation and their practice of sustainable use of diverse resources and identifies the innovative practices as adaptive measures to climate change by the IPLCs that could be brought in to improve the livelihood of the forest dwellers in terms of agriculture and aquaculture by integrating the knowledge of the TRUs to promote, conservation and sustainable use of resources and preserve the biodiversity of the forest.

Chapter 5 presents the traditional knowledge and customary sustainable practices implemented to garner resources from the coastal and marine ecosystems. The rules and resource-using strategies are mentioned for both the TRUs and non-TRUs. In addition, rules followed by other communities have also been discussed. This chapter also discusses shrimp culture and its negative impact on other aquatic species. It demonstrates that traditional knowledge and customary practices of the traditional resource users can promote sustainable fishing practices and are environment friendly as they help in the conservation of marine and coastal resources. It goes as to show that the sustainability and conservation of natural resources depend on the inherent human practices deployed in using such resources and conscious social interaction in tandem with nature promotes natural harmony and thereby ensures conservation.

Chapter 6 explores the multiple values of ecosystem services and their contribution to human well-being in the region. The chapter records the process of shared well-being of both human and nature in terms of mutuality with one another. It also develops the transformative pathways for human–nature relationship. In addition to this, the chapter depicts the current state of the Ecosystem Services (ES) from the TRU's perspective. This chapter also discusses the ecosystem services and the drivers of change in the forest. It demonstrates that both human and nature are dependent on each other and form a human–nature sociality in the ecosystem through which they coexist, therefore promoting human–nature sociality is the key to ensure virtuous cycle in the ecosystem and harmonious relations between human and nature.

Chapter 7 analyses the multiple values of nature and develops a transformational pathway. It explores the TRU's perspectives regarding the services provided by the forest. This chapter further shows an illustration of different meanings of value collated from the TRUs, using IPBES guidelines and demonstrates that market prices do not reckon the true value of natural resources. In addition, price effects and impact of institutional vulnerability on natural resource have been analysed. Finally, three scenarios are discussed on the transformational pathways for sustainable access and benefit-sharing of natural resources.

Chapter 8 shows how the forest acts as a natural barrier to climate change. It explores the impact of climate change, along with the perspective of the TRUs and provides evidence on how the forest has been protecting the people from numerous hazards. It also describes how the forest has proved to be a valuable asset for the country in numerous ways, for all the contributions it has made. The chapter demonstrates that the impact of climate change and natural catastrophe cannot be resisted if humans fail to recognise the perpetual importance of the natural ambience—the Sundarbans and its elements at the coastal belt. It also suggests that current trends of climatic variables, geomorphology, soil nutrients and anthropogenic factors will lead to massive degradation of forest coverage within 2050.

In chapter 9, Community-based Mangrove Agro Aqua Silvi (CMAAS) culture has been thoroughly discussed that is distinct from the commercial rearing of shrimp and other fish resources. The chapter will help understand how CMAAS protects the biodiversity as a substitute to commercial shrimp rearing. The chapter describes the CMAAS culture in detail from its emergence to becoming a cultivating method and a model of farm, etc. The economic impact of CMAAS culture has also been discussed in this chapter. It demonstrates that CMAAS culture following traditional knowledge and customary practices helps protect the biodiversity.

Chapter 10 explores the story of TRUs in the Sundarbans and their struggle towards claiming their right on the forest resources by forming cooperatives. It outlines the steps they have taken and the hurdles they had to overcome to gain their rights. The chapter includes the success stories of three different cooperatives in the Kyora region of Sundarbans and their practices and beliefs. It also illustrates the issues they have realised during the process, the changes they have made, current conditions and the overall outcome. It ushers that forming cooperatives can significantly enhance the resilience capacity of the Indigenous People and Local Communities (IPLCs) including TRUs to combat climate change and other shocks. If IPLCs are empowered, they can significantly conserve and restore the biodiversity of the forest based on their traditional knowledge and customary sustainable practices and reclaim their rights.

Chapter 11 defines the contribution of the traditional resource users of the mangrove forest towards fulfilling Sustainable Development Goals (SDGs) and Aichi Biodiversity Targets. Here, a conceptual framework regarding the resilience and the conservation of the resources of the mangrove ecosystem is shown. It also discusses the political and economic factors that induce biodiversity resource degradation. This chapter recounts the resilience indicators of SEPLS. Moreover, a comparative analysis of the resilience capacity is also illustrated here. It demonstrates that customary sustainable practices and traditional knowledge of the Traditional Resource Users (TRUs) augment conservation through restoration and sustainable uses of biodiversity if TRUs are given a chance and get supported by the government and non-government agencies.

Chapter 12 explores the interdependence between humans and nature and studies the community understanding of 'one health approach' in the face of COVID-19 in the Sundarbans region falling under Bangladesh. The Sundarbans is an ecologically vulnerable area due to the degradation of biodiversity resources triggered by both human interventions and climatic changes. The outbreak of the COVID-19 pandemic and the recent trail of destruction left behind by cyclone *Amphan* have had a catastrophic impact on the forest and millions dependent on the Sundarbans for their livelihood. Against this backdrop, the chapter presents a modified 'one health' framework for recovery post-2020 which will promote human transition into living in harmony with nature.

The framework calls for ensuring right-oriented universal social entitlements, provision of livelihood security and promotion of human–nature cooperation through green production, ecosystem-based approach and biodiverse adaptation to climate change, backed by customary sustainable practices and traditional knowledge in Socio-Ecological Production Landscapes and Seascapes (SEPLS) management.

The concluding chapter that is chapter 13 attempts to outline the key findings of all other chapters. It briefly outlines the conservation criteria of the ecosystem from terrestrial, marine and coastal perspectives; as well as describes traditional rules and practices, multiple values and well-being which are not acknowledged in conducting co-management and conservation process. It therefore demonstrates that the clientelistic network should be broken, and traditional and customary sustainable knowledge should be considered as a vital part of the co-management process. In addition, it ascertains that the indigenous and local knowledge needs to be integrated, while formulating and implementing rules, policies and strategies for the conservation, restoration and preservation of the forest in the face of massive biodiversity degradation. The chapter also illustrates the importance of promoting a whole of the society approach bringing all under the umbrella practices of sustainable production and consumption as well as equal access and benefit-sharing for the post-2020 period.

We would not be able to undertake this research without the contribution and participation of the members of three cooperatives of Traditional Resource Users (TRUs) in Koyra Upazilla of Khulna district—*Koyra Bonojibi Bohumukhi Unnayan Samity*, *Harinagar Bonojibi Bohumukhi Unnayan Samity* and *Munda Adivasi Bonojibi Bohumukhi Unnayan Samity*. They have been with us throughout the research and provided valuable information. We are grateful to them.

Since the beginning of activities, the Forest Peoples Programme (FPP), UK, have been providing financial and technical support. Without their continuous support and guidance, particularly from Caroline de Jong and Maurizio Farhan Ferrari, we would not be able to do this work. We are grateful to the FPP team.

Dr. Tanjinul Haque Mollah was involved in a part of this work with technical assistance. He and his team have provided overall support to conduct PPGIS. I thank him too.

We got logistic, HR, technical and financial supports from *Unnayan Onneshan*. Many of the materials of research are from *Unnayan*

Onneshan's data reservoir. In particular, Azmol Hossain and Mamun-or-Rashid of *Unnayan Onneshan* have been involved in this work from the very beginning and have assisted in various ways throughout the research. It would not have been possible to write the chapters without their help.

Finally, I would like to thank all of the authors for their contributions, and continuous engagement and interactions as part of the community practice of *Unnayan Onneshan*.

Dhaka, Bangladesh Rashed Al Mahmud Titumir

Contents

Editor and Contributors

About the Editor

Rashed Al Mahmud Titumir is a Professor of economics at the Department of Development Studies at the University of Dhaka and Chairperson of the Unnayan Onneshan, a centre for research and action on development in Dhaka, Bangladesh. He is currently holding the position of Vice-Chairperson of IUCN Asia Regional Members Committee and Chairperson of IUCN National Committee of Bangladesh.

Contributors

Afrin Tanjila is an Assistant Professor at the Department of Development Studies at Bangladesh University of Professionals and also an Adjunct Researcher at Unnayan Onneshan.

Islam Mohammad Saeed is an Assistant Professor at the Department of Development Studies at Bangladesh University of Professionals, and also an Adjunct Researcher at Unnayan Onneshan.

Meem Mahtarin Haq is a Research Intern at Unnayan Onneshan.

Paran Md. Shah is a Researcher at Unnayan Onneshan.

Pasha Mostafa Walid is a Researcher at Unnayan Onneshan.

Acronyms

ADB	Asian Development Bank
BBS	Bangladesh Bureau of Statistics
BCR	Benefit-Cost Ratio
BFD	Bangladesh Forest Department
BMD	Bangladesh Meteorological Department
CEGIS	Center for Environmental and Geographic Information Services
CMAAS	Community-based Mangrove Agro Aqua Silviculture
CS	Commercial Shrimp Culture
DoE	Department of Environment
DPSIR	Driver, Pressure, State, Impact and Response
ECA	Ecologically Critical Area
FAO	Food and Agriculture Organization
FD	Forest Department
FGD	Focus Group Discussion
GBM	Ganges, Brahmaputra and Meghna
GDP	Gross Domestic Product
GEF	Global Environment Facility
GIS	Geographic Information System
GoB	Government of Bangladesh
ICZMP	Integrated Coastal Zone Management Project
ILK	Indigenous and Local Knowledge
IPLCs	Indigenous People and Local Communities
IUCN	International Union for Conservation of Nature
MSY	Maximum Sustainable Yield
NB	Net Benefit
NDF	Netherlands Development Funds

NGOs	Non-Governmental Organizations
NPV	Net Present Value
NWFPs	Non-Wood Forest Products
PA	Protected Area
PNG	Papua New Guinea
PPGIS	Public Participation Geographic Information System
PVB	Present Value of Benefit
PVC	Present Value of Cost
RF	Reserve Forest
SBCP	Sundarbans Biodiversity Conservation Project
SDGs	Sustainable Developments Goals
SEPLS	Socio-Ecological Production Landscapes and Seascapes
SRF	Sundarbans Reserve Forest
TEK	Traditional Ecological Knowledge
TRUs	Traditional Resource Users
UNESCO	United Nations Educational, Scientific and Cultural Organization
UNU- IAS	United Nations University Institute for the Advanced Study of Sustainability
UO	Unnayan Onneshan
WB	World Bank

List of Figures

List of Tables

CHAPTER 1

Introduction: The Sundarbans as Our Mind

Rashed Al Mahmud Titumir

Introduction

The Sundarbans is located in the largest delta of the Ganges, Brahmaputra and Meghna (GBM) rivers before these waterbodies enter the Bay of Bengal leaving a humongous serpentine trail behind. Sundarbans, a local connotation for the abode of Sundari tree boasts of being the largest neighbouring single-tract contiguous mangrove ecosystem in the world (Fig. 1.1). It is nature's symbol of beauty, tranquillity and a biodiversity hotspot.

Geographically, this largest mangrove forest is located in the southwest corner of Bangladesh, between 21°30′ and 22°30′ North and 89°00′ and 89°55′ East (Islam, 2010). This mangrove forest is not only confined to Bangladesh but a significant part of it lies within India as well in the state of West Bengal. However, the area lying inside Bangladesh is larger in

R. A. M. Titumir (✉)
Department of Development Studies, University of Dhaka, Dhaka, Bangladesh
e-mail: rt@du.ac.bd; rtitumir@unnayan.org

Unnayan Onneshan, Dhaka, Bangladesh

R. A. M. Titumir (ed.), *Sundarbans and its Ecosystem Services*, Sustainable Development Goals Series,
https://doi.org/10.1007/978-981-19-3000-3_1

area as compared to the share falling in India. The area comprises 6,071 square kilometres (62% of total area) and constitutes 39.5% of the total forest area of Bangladesh (Roy & Alam, 2012).

The part of this mangrove forest which falls under Bangladesh, consists of 70% landmass and 30% water (Kabir & Hossain, 2008). The wetlands of this mangrove forest consist of about 200 islands, which are separated by about 400 interconnected tidal rivers, creeks and canals (Rahman et al., 2010). Moreover, this forest was recognised as a natural World Heritage site (139,700 hectares of forest land comprising Sundarbans East, West and South) in 1997 by UNESCO. It was considered as a Ramsar Site of international importance in 1992 (IUCN, 2014). It was also included on the list for the selection of the 'Seven Wonders of the World'.

A vast majority of people living in this area depend on the forest for their livelihoods by utilising its resources that make it a unique hotspot for biodiversity conservation. The area of Sundarbans that falls within the geographical boundaries of Bangladesh considered as the lung of the country is currently identified as an eco-sensitive area that has been left vulnerable due to the constant degradation of its natural resources impacting its unique biodiversity. Over the past few decades, the area has witnessed major ecological and physiographical changes resulting in a major depletion of its resources caused by both human interference and natural calamities triggered due to climate change. These changes have had a damaging impact on the mangrove forest, one among few left in the world. The major anthropogenic causes of degradation include illicit destruction, conversion of forestland into agriculture and aquaculture for shrimp cultivation, etc.

This chapter comprises the description of the biodiversity of this mangrove forest along with an emphasis on different kinds of flora and fauna and resource systems. Apart from it, an alternative conceptualisation of the values by the Traditional Resource Users (TRUs) has been developed here. In addition to it, the chapter depicts the current management system in place along with the problems associated with it. Lastly, it mentions the methodology of the study adopted throughout the book, briefly describes the rationale and scope of the book, and provides a sneak peek of the upcoming chapters and the interesting read that the subsequent chapters have in store.

Biodiversity Resources

The Sundarbans is an amalgamation of three different ecosystems namely forest, coastal and wetland which allow this forest to house a variety of aquatic and terrestrial flora and fauna flourishing in the area by adapting uniquely to its surrounding. It is unique as it supports viviparous plant species and harbours 334 species of trees, shrubs, herbs and epiphytes and around 400 species of wild animals (Behera & Haider, 2012). Considered a biodiversity hotspot, it harbours a variety of flora and fauna in the coastal and marine regions. The distribution of diverse species of flora and fauna depends on the habitats' features. The habitats' conditions differ as per the spatial and temporal variability of hydrological regimes. Additionally, reports show that the inflow of freshwater and the tidal variation, the salinity and the topology of different areas have built a highly heterogeneous habitat condition for this mangrove forest (Gopal & Chauhan, 2006).

Floral Species

A total of 65 floral species consisting of 37 families have been reported in the Bangladesh portion. Among which, 10 are large trees, 20 are small trees, 25 are shrubs, 10 are herbs and two are ferns. However, a recent study revealed 115 floral species that showed a significant expansion in floral diversity. Apart from 10 large trees, this study adds seven more trees, ten species of sedges, five species of grasses and many other species from various groups. It is also reported that a total of 35 taxa of Phytoplankton communities including nano and picoplankton exist. This number further extends to 50 if freshwater phytoplankton is reported that is dominant under all salinity conditions. There are a total of 35 benthic algae.

Among 50 true mangrove plant species recorded throughout the globe, this forest alone contains 35 species (Rahman & Asaduzzaman, 2010). The Sundari (*Heritiera fomes*) is the most climax tree species. Other prominent species include: gewa (*Excoecaria agallocha*), baen (*Avicinnia officinalis*), passur (*Xylocarpus mekongensis*), keora (*Sonneratia apetala*), goran (*Ceriops decandra*), ora (*S. caseolaris*) and hental (*Phoenix paludosa*). It also offers high value non-timber forest products like golpata (*Nypa fruticans*). There are about 13 and 23 species of orchards and medicinal plants in this forest.

Faunal Species

This region is also rich in its faunal diversity. There are 448 species of vertebrates that include 10 amphibians, 58 reptiles, 339 birds and 41 mammals (DoE, 2015). Another study suggests the forest contains 1,136 species of wildlife that includes 315 species of birds (84 migratory), 678 aquatic species (210 fish species), 59 reptiles, 8 amphibians, 11 cetaceans and 16 mollusks (Aziz & Paul, 2015).

It provides habitats for diverse aquatic wildlife such as the estuarine crocodile (*Crocodylus porosus*), turtles (*Lepidochelys olevacea*), dolphins (*Platanista gangetica* and *Peponocephala electra*) and molluscs like the giant oyster (*Crassostrea gigas*). The rich avifauna of the forest includes mangrove pitta (*Pitta megarhyncha*), mangrove whistler (*Pachycephala grisola*), brown-winged kingfisher (*Halcyon amauroptera*) and collared kingfisher (*Todiramphus chloris*) (Khan, 2005). Other various species are purple heron, pond heron, cattle egret, little egret, spotted dove, tailor birds, magpie robin, woodpeckers, barbets, bee-eaters, bulbuls, shrikes, starlings, babblers, thrushes, orioles, flycatchers, etc. (Kabir & Hossain, 2008; Rahman & Asaduzzaman, 2010).

Among all faunal species, the most magnificent animal is the Royal Bengal Tiger (*Panthera tigris*). According to the census of 2004, around 440 tigers are settled in Bangladesh. However, the most recent reports estimated the existence of around 114 tigers (Bangladesh Forest Department [BFD], 2019). This forest is also home to numerous spotted deer (*Axis axis*) and barking deer (*Muntiacus muntjak*). Other animal species include wild boars (*Sus scrofa*), monkeys, jungle cats (*Felischaus*), rhesus macaque (*Macaca mulata*), otters (*Lutrapers picillata*), fishing cat, civet cat, Bengal fox and jackal. Furthermore, it is known to be the largest honey-producing habitat in the country with giant honey bees (*Apis dorsata*).

Aquatic Resources

The Sundarbans also has endless aquatic resources as it occupies an extensive part of the marine and coastal zone with around 1,874 square kilometres of the water body. In addition, it includes a 20 km wide marine zone that is offshore fishing area that covers an area of 1,603 square kilometres (Haque, n.d.). The coastal and marine belt of the forest occupies diversified fish fauna varieties including 53 pelagic and 124 demersal

Fig. 1.1 Mangroves in the Sundarbans (*Photo* Unnayan Onneshan)

species (Sarkar, 1989; Rainboth, 1991 cited in Gopal & Chauhan, 2006). Nonetheless, the distribution of fishery resources can be categorised as per salinity gradients.

The saline area contains fish like *Harpodon nehereus*, *Trichiuruss avala*, *Setipinna* sp., *Pampus* sp., *Sardinella* sp. and *Salar* sp. On the other hand, freshwater or low salinity areas are full of *Pangasius pangasius* and *Lates calcarifer*. The moderately saline water or brackish water zones contain *Hilsa* (*Tenualosa*) *ilisha*, *Pomadasys hasta*, *Polynemus* sp., *Coilia* sp., *Paissha* (*Setipinnaphasa*), *dadne*, *vetki* (*Latescal carifer*), *kaine*, *tengra* (*Mystus tengara*), *ghagra*, *baila* (*Periophthal modonschlosseri*) and others (Islam and Haque, 2004 cited in Gopal & Chauhan, 2006). Likewise, the waters in the forest also harbour Ganges River dolphins, Irrawady dolphins, Indo-Pacific humpback dolphins and finless porpoises. In general, fishermen in the forest catch approximately 120 species of fish. There are eight commercially important breeds including finfish, shrimp and crabs. The estuarine zones also harbour at least 11 species of bivalves (Aziz & Paul, 2015).

As for the invertebrates, there are 20 species of prawns, 8 species of lobsters, 7 species of crabs, several species of gastropods and 6 species of pelecypods (Kabir & Hossain, 2008). The yearly production of the marine zone has been estimated at 8,733 metric tonnes or 54.5 kg/ha (Haque, n.d.). Fishing in the area is largely influenced by its climate and from May to August, fishing is regarded as a hazardous and risky activity. In addition, the weather also turns rough due to cyclones and storms causing floods and other natural calamities.

Resource System

Both biotic and abiotic resources contribute directly or indirectly to the economy at local and national levels. These resources have been used for various purposes, contributing both to the lives and livelihoods of local people and the economy of the country (Fig. 1.2). Wood and golpata (*Nypa fruticans*) collectors (Bawalis), fishermen (Jele), honey and wax collectors (Mouals), shell collectors (Chunary) and crab collectors are among the major occupational groups found in the region adjacent to the forest. The lives and livelihoods of these local people are mostly related to the physical and biological (or biodiversity) resources (Fig. 1.2).

The resource system can be classified into six parts (Fig. 1.2). These are physical, chemical, ecological, aesthetic and recreational categorised as biological and medicinal. The physical resources consist of soil (used for agricultural, organic fertiliser and land stabilisation) and water (used for irrigation, fisheries and navigation). Then there are chemical resources that include glues and dyes, potassium carbonate, salt, alcohol and acetic acid, wood tar, oil and tar oils derived from various elements of the forest. Ecological services are of 4 types; these are nutrient cycle, carbon storage, natural protection from cyclone and storm surge and habitat provision (consisting of different kinds of flora, birds and animals). The aesthetic and recreational services consist of wildlife viewing, nature walks and world heritage and Ramsar sites. The biological resources include plants (used for boat building, fodder, food, stakes for fishnets, straps, etc.) and animals (used for their skin, honey and wax, food, etc.). Lastly, the medical resources are used for rheumatic disorder and anti-diarrheal purposes.

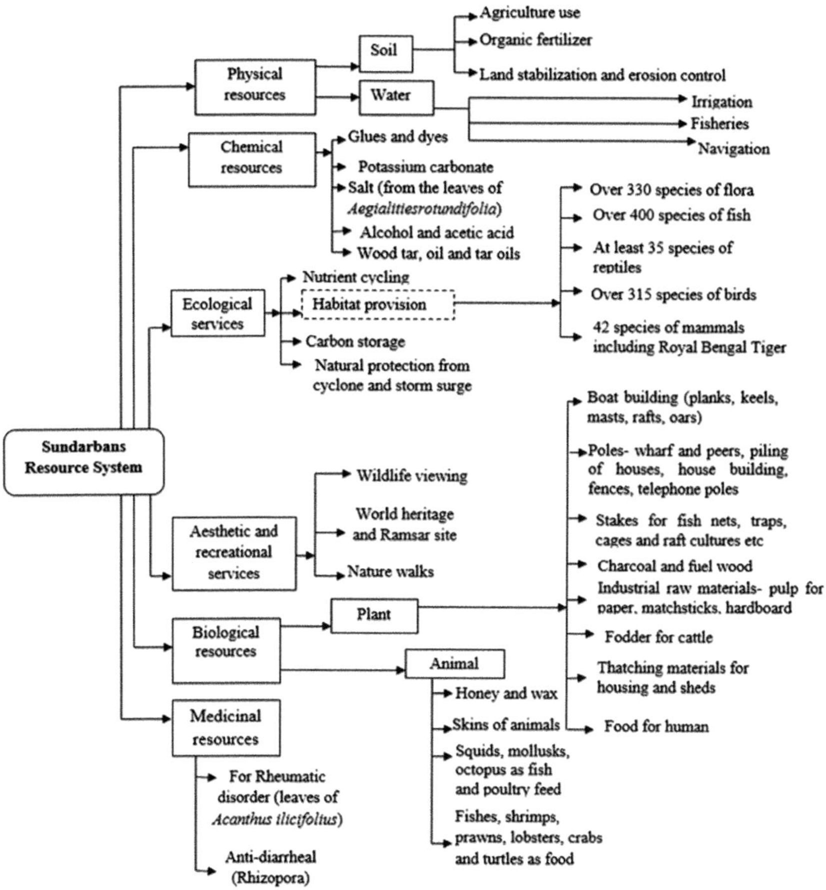

Fig. 1.2 The resource system (*Source* Adapted from data reservoir of *Unnayan Onneshan* as cited in Titumir & Afrin, 2017)

An Alternative Conceptualisation of Values by the TRUs

The TRUs consider the Sundarbans to be their mind (Titumir & Afrin, 2017; Titumir & Paran, 2022a; Titumir et al., 2019). They equate their life and well-being with the life and well-being of the forest as well as their thinking process and the way of life around the life and spirit of the forest (Titumir & Paran, 2022a, 2022b, 2022c). It looks into the people's existence, their social harmony, the breeding of their offspring, natural safety and security. Human beings count on it. Their lives are influenced by the services offered by this forest which combines many values such as use values and non-use values.

The following Fig. 1.3 shows how the forest is considered to be "our mind" and how it is interlinked with the ecosystem services (provisioning, regulating and supporting). This link-up can be seen by the direction of the arrows in the diagram. Between the two kinds of values, use value consists of direct use value (food, fuel and water), indirect use value (climate regulation, cyclone protection and erosion control) and option value (research, education and aesthetics). The non-use value consists of bequest value (cultural heritage and historical legacy) and existence value (biodiversity and habitat).

Each kind of value has a particular meaning to the TRUs. They classify the Ecosystem Services (ES) based on their age-old dependence and traditional knowledge system. Their conceptualisation of multiple values of the ES can be collated with the scientific understanding of values. Direct use value means the economic or social value of the goods or benefits derived from the services provided by an ecosystem that are used directly by an economic agent. These include consumptive uses (for example, harvesting goods) and non-consumptive uses (for example, enjoyment of scenic beauty). Indirect use value includes benefits derived from the goods and services provided by an ecosystem that are used indirectly by an economic agent. For example, indirect use values are the provision of cyclone protection by the forest or the usage of natural water of the forest, etc. Existence value is often reflected as a sense of well-being such as the existence of coastal and marine biodiversity and habitats. People may not have experienced or utilised this value directly but they benefit from the knowledge of it. Option value refers to the value of retaining options for the future. These values include the potentials of biodiversity that are presently unknown and need to be explored. It is the value of knowing that biological resources exist in this biosphere that may prove to be an effective option for something important in the future. For example, research helps explore the diversity of myriad flora and fauna of the forest. Education and aesthetics are also examples of option value. Bequest value refers to the readiness of present-day mankind to spend goods in order to preserve biological diversity and its components for future generations. Bequest value is often termed as beneficial or altruistic value. For example,

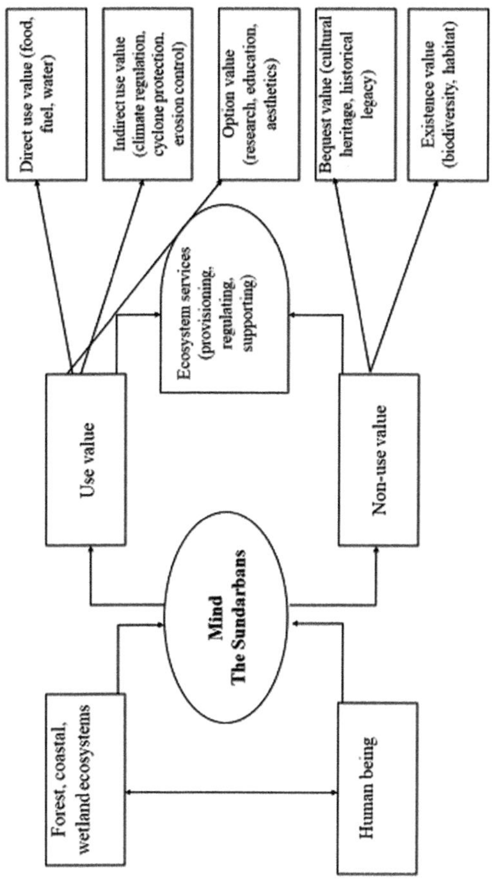

Fig. 1.3 The Sundarbans as our mind (*Source* Adapted from Titumir et al., 2019)

cultural heritage and resources for future generations are also expressed through consumers' willingness to pay or less/delayed consumption of the resources for the sake of the future.

Current Management Approach and Associated Problems

The forest was first announced to be a Reserved Forest (RF) during British rule. In the post-independence period, after 1971, it was once again declared as an RF under the Forest Act of 1927 that restricted any activity within the forest boundaries without prior formal permission. Hence, the key management right has been under government control for long. Later, the forest policy of 1994 acknowledged community participation in the management process and recognised certain rights of the local people. Thus, the property rights structure within the periphery of the forest cannot be categorised under a specific type as in the case of private and public property. In contrast, the right to property within the periphery of the forest is equally distributed among the state and local people. The long-term framework of property rights can be demonstrated by a diagrammatic representation based on Schlager and Ostrom's (1992) typology of a bundle of property rights (Fig. 1.4).

Figure 1.4 shows the property rights structure where three categories can be seen from the state before and after 1994. Before 1994, these were classified as proprietor, authorised claimant and authorised user. After 1994 these equivalently changed to exclusion, management and access

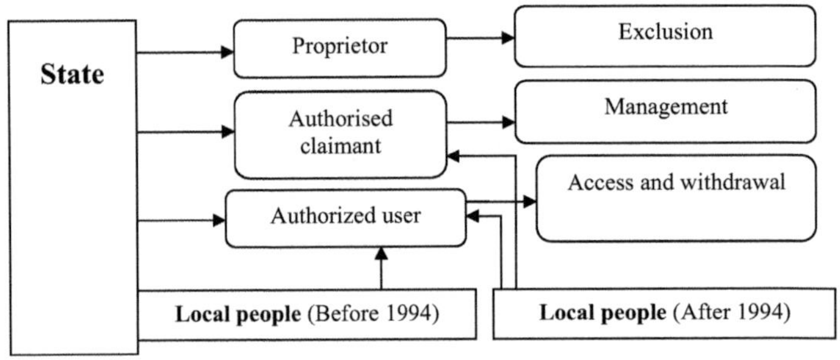

Fig. 1.4 Property rights structure (*Source* Prepared by author based on Schlager & Ostrom, 1992)

and withdrawal. The local people were always the authorised user both before and after 1994. However, after 1994 they also became authorised claimant.

The government has taken the forest under its control, to conserve its resource and save it from being overexploited by the people. The state authority, particularly the Forest Department (FD), has taken the responsibility to assure that resources are used efficiently. The TRUs have the right to access and use resources only with the permission from FD. Furthermore, as mentioned earlier, the local people were only given the rights to access and withdrawal before the adoption of the 1994 Forest Policy. From 1994, the local people have been recognised as rightful claimants of the forest resources at the policy level. This shows that they have management rights alongside their access and withdrawal rights. According to the policy, under a cooperative mechanism, the people should be consulted in terms of determining the management policy of the forest resources. This is because they have the right to regulate the internal usage pattern of the resources. This implies the formal institutional arrangement to be unstable. Local people are barely involved in the decision-making process. They face many hurdles to exercise their rights, to have access inside the forest and to harvest the resources. In contrast, the state authority (FD) has been exercising the rights as owner, proprietor, authorised claimant and other authorised users. The conservation procedure of the resources, under this management framework, turned out to be unproductive. As a result, the management approach greatly misses to cease the degradation process of the resources (see Titumir et al., 2020, 2022).

Methodology

The studies in this book are carried out using the Multiple Evidence Base (MEB) approach. A significant amount of data was collected through participatory observations, Focus Group Discussions (FGDs), and Key Informant Interview (KII) in three cooperatives. It particularly draws on the Indigenous and Local Knowledge (ILK) of the forest people of *Harinagar Bonojibi Bohumukhi Unnayan Samity*, *Koyra Bonojibi Bohumukhi Unnayan Samity* and *Munda Adivasi Bonojibi Bohumukhi Unnayan Samity* that the Unnayan Onneshan helped set up. Members of the cooperatives pursue their livelihoods as wood collectors (Bawalis), fishermen (Jele), honey and wax collectors (Mouals) and crab collectors. Focus group discussions were conducted based on these traditional professions of the forest people. It also conducted KII with unstructured questionnaires. Besides, it adopts PRA tools such as resource mapping, impact analysis by the members of cooperatives.

It uses the data reservoir of the Unnayan Onneshan research institute that initiated several biodiversity conservation programmes and carried out research on the Sundarbans. The study also collected data using the Public Participation Geographic Information System (PPGIS) to look into the decadal changes of ES, landscape and ecological valuation, climate impacts. Results from FGDs and PPGIS have been cross-checked through supporting literatures. The study therefore links numerous sources of scientific knowledge to bring forth a comprehensive and scientific understanding of the ecosystem in the Sundarbans.

PPGIS Method

Several Remote Sensing Imageries are used in this study where two Landsat images were downloaded from the United States Geological Survey (USGS) and one sentinel-2 image from European Commission's Copernicus programme. Two Landsat 5 Thematic Mapper (TM) for the analysis of 2000, 2010 and One Sentinel-2 Multi-Spectral Instrument (MSI) were downloaded for 2020. Assessment processed by Erdas Imagine 2014, ArcGIS 10.5 and Google Earth Pro. The characteristics of the satellite image used in this study is shown in Table 1.1.

In this study, each satellite image was classified in three classes using Maximum Likelihood Classification in ArcGIS listed in Table 1.2 with their training sample as well as ground control points. Maximum Likelihood is a Supervised Classification where a set of training samples is applied as input. It is assigned to indicate a class that has the probability of

Table 1.1 Detailed information of satellite images used for PPGIS

Satellite imagery	*Path/row*	*Acquisition date*	*Resolution (m)*
Landsat 5 Thematic Mapper (TM)	137/045	19/01/2000	30
Landsat 5 Thematic Mapper (TM)	137/045	30/01/2010	30
Sentinel-2 MultiSpectral Instrument (MSI)	137/045	01/02/2020	10

Source Unnayan Onneshan (2020)

Table 1.2 Based on LULC each satellite image was in the following categories

Class	*Training sample × 9 pixels*	*Ground Control Points (GCP)*
High density	10	4
Medium density	15	9
Low density	18	12

Source Unnayan Onneshan (2020)

being that assigned class in the real world and ground control points are used to validate the classification performance. All ground control points were collected in September 2020. For landscape value mapping of TRUs kernel density is used to estimate the probability density function.

Study Area

The area selected for the study is 40 km in length and 30 km in width. The region is located between 22°28′30″ north to 22°1′0″ north latitude and 89°13′30″ east to 89°30′0″ east longitude. This region is part of the Khulna Range—one of the four administrative areas of the Sundarbans.[1] On the west of the region is Koyra upazila of Khulna district and the Satkhira Range of the Sundarbans. On the east, the Shibsa river separates this area from the Dacope upazila of the Khulna Range and the Chandpai Range. The Rai river separates Koyra Upazila from the Sundarbans. The locals call the river Rai as Shakhbaria river. It is one of the numerous tributaries of the Sundarbans. The Rai River goes further south and joins the Arpangasia river in the Bay of Bengal. There is a complex net of waterways as numerous rivers, tributaries and canals spread across the Sundarbans. Local people call them by different names.

Rationale and Scope of the Book

The book is an interdisciplinary inquiry undertaken in the Sundarbans, which brings Ecology, Anthropology, Sociology, Economics and Political Economy in a shared platform. Though there are attempts to study the Sundarbans from different disciplines separately, there is an absence of

[1] The other three administrative areas of Sundarbans are: Chandpai Range, Sarankhola Range and Satkhira Range.

comprehensive understanding adopting multiple evidence-based approach that includes both primary and secondary data and brings heterodox multidisciplinary perspectives in a common platform. Hence, this book adopts Multiple Evidence Base (MEB) approach in order to bring participatory insights of the Indigenous People and Local Communities (IPLCs) including TRUs.

Firstly, it presents a comprehensive story of the Sundarbans—a natural World Heritage Site and a Ramsar Site of international importance and introduces a new theoretical framework by combining theory, praxis and policy in a triangulation method and advocates for ensuring rights-oriented universal social entitlements, provision of livelihood security and promotion of human–nature cooperation to ensure the well-being of both human and nature.

Secondly, the chapters present a scientific narrative of the majestic mangrove forest to both researchers and general readers, and is presented in a way that helps understand any mangrove forest or composite ecosystems across the world.

Thirdly, it challenges the market-based approach in valuing the natural resources and demonstrates that valuation of environmental resources through market penetration pricing does not reckon the social benefits and values coproduced through complementarity between humans and nature.

Fourthly, it brings the institutional vulnerability aspects and power-sharing arrangements into the analysis of resource management and argues that due to lack of a proper market structure and equal distribution of power, rents are dissipated through market pricing, going into the pockets of the rent-seeking powerful class. This rent-seeking behaviour induces unproductive, expropriating activities that bring positive returns to the individual but not to society.

Fifthly, it formulates a new understanding of sustainability in case of usage and management of natural resources by incorporating the idea of human–nature sociality that accounts for the interdependent relationship between TRUs and the Sundarbans ascertaining the living in harmony with nature.

The chapters in the book provide different perspectives that are distinctive from conventional understanding for several reasons. First, the book advocates for promoting human–nature cooperation to maintain ecosystem health. This framework highlights the problems in valuing natural resources from market centric approach. It precisely contributes to the sphere of knowledge by recognising that a complementary solution to natural resource management can be organised through a mutual relation between human and nature against the mainstream theories where self-interest maximisation creates a rupture in the ecosystem. In contrast to non-sustainable use of the resources (expropriation), this framework demonstrates that proper appropriation of the natural resources is healthy and can ensure well-being of both human and nature. The human–nature bidirectional causality, established throughout the book, entails that both are dependent on each other. Furthermore, the chapters demonstrate that the socially constructed norms and values which are the informal institutional set up can solve the natural resource problem. Second, the chapters are written in a holistic and interdisciplinary way. As a result, they offer a comprehensive picture of the Sundarbans to the readers combining theory, praxis and policy in a triangulation method. Hence, this book provides for on the ground learning for evidence-based policymaking and meets the academic interest.

The book can potentially contribute to the upcoming nexus assessment of Intergovernmental Science-Policy Platform on Biodiversity and Ecosystem Services (IPBES) and the Convention on Biological Diversity (CBD) post-2020 global biodiversity framework through providing useful knowledge and information on Socio-Ecological Production Landscapes and Seascapes (SEPLS) approaches to sustainable management of natural resources.

It can also assert tangible contribution to Sustainable Development Goals (SDGs—1, 2, 3, 4, 5, 8, 11, 12, 13, 14, 15 and 16) and Education for Sustainable Development (ESD) for 2030 priority action areas (1, 2, 3, 4, 5) for achieving good health and well-being of both ecosystem and human being within the SEPLS.

It will be useful to equip students, academicians, researchers and scholars involved in interdisciplinary learning and research with the acumen to understand the complexities of natural resource governance in developing countries. Although the narration style of the book is simple and fluent, the language rigour is maintained. It ushers a scientific narrative of the majestic mangrove forest to both researchers and general readers, and is presented in a way that helps understand any mangrove forest or composite ecosystems across the world.

Design of the Chapters

This book contains thirteen chapters. Chapter 2 explains the degradation manner of the resources at present, in the terrestrial ecosystems that is done by developing a DPSIR framework and interpreting current management practices. Chapter 3 discusses the present condition of floral and faunal species in the coastal and marine ecosystem and shows how the marine ecology is being disrupted due to the expropriation of resources. Chapter 4 focuses on how traditional rules and practices upheld by the TRUs in the terrestrial region of the forest instinctively lead to the conservation of the forest's ecosystem and biodiversity. Chapter 5 explains the traditional rules and practices in marine and coastal zones where traditional resource users utilise their age-old conservational practices while gathering resources from that belt. It also shows how clientelism is harming the aquatic resources of the marine and coastal ecosystem of the forest. Chapter 6 reveals the forest's ecosystem services and their values—exploring how the services are beneficial to the well-being of both humans and nature. Chapter 7 demonstrates multiple value assessments of nature that criticises the orthodox monetary school of thought, about the valuation of nature and proposes an alternative approach. Chapter 8 examines the fearless role played by the Sundarbans in response to climatic degradation. It also suggests that the forest works as a natural shield against natural disasters, carbon emission, sea level rise, tidal waves, erosion, etc. Chapter 9 discusses the Community Based Mangrove Agro Aqua Silvi (CMAAS) Culture. This is a new approach practised within the forest people communities as a part of their conservation criteria. Chapter 10 attempts to show the case of three forest people cooperatives, who deal with various socio-political hurdles, in the way of achieving and claiming their rights. It also combines many case studies from a specific region to reveal how the cooperative members have developed during the years. In Chapter 11, a new sustainable conservation framework, to boost the resilience capacity of the forest ecosystem has been introduced. It also defines in detail the contribution of the traditional resource users of the mangrove forest in achieving the Sustainable Development Goals (SDGs) and Aichi Biodiversity Targets. Chapter 12 explores the interdependence between humans and nature and studies the community understanding of 'one health approach' in the face of COVID-19 in the Sundarbans region. Finally, Chapter 13 summarises the entire discussion. It wraps up the entire book by recalling parts of every chapter and briefly collaborating all of it.

Conclusions

This chapter helps to provide an introduction to the book and its content, in addition to a brief summary of each segment. It explains what exactly the largest mangrove forest is about and what its inhabitants consider it to be. It starts by providing a glimpse of the geographic and physiological features of the mangrove forest. The chapter opens with the location of the forest, as to where it is situated to what area it covers; an entire scenario is drawn in the beginning. Additionally, it goes on to talk about the various kinds of resources along with their classifications that include the flora and fauna including different kinds of plants, animals and marine species. Besides, different kinds of fish as resources were also commented upon. The classification of species and what they have to offer, their contribution to the ecosystem, biodiversity, etc., have been illustrated. These different varieties reflect how unique and distinct each of these species is. It also helps to reflect the purpose of each of these resources, in terms of what it has to offer to humans, as well as to nature. A breakdown of the resource system was shown as well. Moreover, both the biotic and abiotic resources of the forest and their contribution to the economy at local and national levels have been discussed as well.

A brief description has been included about the alternative conceptualisation of the values adopted by the TRUs alongside their perspective. These values ensure present existence and the interlinking of cordial bonds among the forest and people alongside goods and services for the upcoming generations. The values are mainly of two types, that is, direct and indirect. Direct values consist of supplies such as food, while indirect value takes into account the likes of climate regulation, cyclone protection, erosion control and option values such as research, education and aesthetics. Furthermore, the chapter analysed how the forest forms an orientation of 'being the mind' to the people. It symbolises the harmony that humans ought to share with the forest. The chapter also revealed that humans are dependent upon this mangrove ecosystem and its resources. In addition to these, several examples of direct, indirect and option use values are also provided here. Also, what these values are, how they are connected to nature and humans, what they have to offer and so on, are described in this chapter. The current management issue and existing problems are also stated here. This includes the property rights structure and how it is causing an effect. It also depicts that the formal institutional arrangement is unstable; along with the kind of rights, access and authority the common people and TRUs have.

Acknowledgements Many of the insights of the chapter are drawn from 'The Sundarbans is our mind: an exploration into multiple values of nature in conversation with traditional resource users', published by United Nations University Institute for the Advanced Study of Sustainability, Tokyo, Japan, 2019, in *Understanding the multiple values associated with sustainable use in socio-ecological production landscapes and seascapes (SEPLS)*, Satoyama Initiative Thematic Review, Vol. 5.

References

Aziz, A., & Paul, A. R. (2015). Bangladesh Sundarbans: Present status of the environment and biota. *Diversity, 7*(3), 242–269.

Bangladesh Forest Department [BFD]. (2019). *Status of tiger in Bangladesh Sundarbans 2018: Second phase*. Bangladesh Forest Department. Ministry of Environment and Forests, Government of the People's Republic of Bangladesh.

Behera, M. D., & Haider, M. S. (2012). *Situation analysis on biodiversity conservation—Ecosystem for life—A Bangladesh India initiative*. IUCN.

DoE. (2015). *Fifth national report to the convention on biological diversity: Biodiversity National Assessment and Programme of Action 2020*. Ministry of Environment and Forest, Government of the People's Republic of Bangladesh, Dhaka, Bangladesh.

Gopal, B., & Chauhan, M. (2006). Biodiversity and its conservation in the Sundarbans mangrove ecosystem. *Aquatic Sciences, 68*, 338–354.

Haque, M. D. E. (n.d.). *How fishers' endeavours and information help in managing the fisheries resources of the Sundarban mangrove forest of Bangladesh* (pp. 433–438). Bangladesh Forest Department.

Islam, S. N. (2010). Threatened wetlands and ecologically sensitive ecosystems management in Bangladesh. *Frontiers in Earth Science, 4*(4), 438–448.

IUCN. (2014). *Bangladesh Sundarban delta vision 2050: A first step in its formulation*. Bangladesh Country Office, Dhaka, Bangladesh.

Kabir, D. M. H., & Hossain, J. (2008). *Resuscitating the Sundarbans: Customary use of biodiversity and traditional cultural practices in Bangladesh*. Unnyan Onneshan, BELA, Forest peoples Programme and Nijera Kori.

Khan, M. M. (2005). Species diversity, relative abundance and habitat use of the birds in the Sundarbans East Wildlife Sanctuary, Bangladesh. *Forktail, 21*, 79–86.

Rahman, M. R., & Asaduzzaman, M. (2010). Ecology of Sundarban, Bangladesh. *Journal of Science Foundation, 8*(1&2), 35–47.

Rahman, M. M., Rahman, M. M., & Islam, K. S. (2010). The causes of deterioration of Sundarbans mangrove forest ecosystem of Bangladesh: Conservation and sustainable management issues. *Aquaculture, Aquarium, Conservation & Legislation International Journal of the Bioflux Society, 3*(2), 77–90.

Roy, A. K. D., & Alam, K. (2012). Participatory forest management for the sustainable management of the Sundarbans mangrove forest. *American Journal of Environmental Science, 8*(5), 549–555.

Sarkar, S. U. (1989). Fish eating wildlife and some fishes of the Sundarbans, Bangladesh. *The Journal of NOAMI, 6*(1 & 2), 17–29.

Schlager, E., & Ostrom, E. (1992). Property rights regimes and natural resources: A conceptual analysis. *Land Economics, 68*(3), 249–262.

Titumir, R. A. M., & Afrin, T. (2017). Complementarities of human-nature well-beings: A case illustrated through traditional forest resource users of Sundarbans in Bangladesh. In *Satoyama initiative thematic review (SITR)—Volume 3.* United Nations University Institute for the Advanced Study of Sustainability (UNU-IAS).

Titumir, R. A. M., Afrin, T., & Islam, M. S. (2020). Traditional knowledge, institutions and human sociality in sustainable use and conservation of biodiversity of the Sundarbans of Bangladesh. In O. Saito, S.M. Subramanian, H. Hashimoto & K. Takeuchi (Eds.), *Managing socio-ecological production landscapes and seascapes for sustainable communities in Asia: Mapping and navigating stakeholders, policy and action* (pp. 67–92). Science for Sustainable Societies. https://doi.org/10.1007/978-981-15-1133-2_5

Titumir, R. A. M., Afrin, T., & Islam, M. S. (2022). *Natural resource degradation and human-nature wellbeing: Cases of biodiversity resources, water resources, and climate change.* Springer (Forthcoming).

Titumir, R. A. M., & Paran, M. S. (2022a). Human-nature cooperation for well-being: Community understanding on one health approach in the COVID-19 era in the Sundarbans. In M. Nishi, S. M. Subramanian, & H. Gupta (Eds.), *Biodiversity-health-sustainability nexus in socio-ecological production landscapes and seascapes (SEPLS).* Springer Nature.

Titumir, R. A. M., & Paran, M. S. (2022b). Ecosystem services and well-being in the Sundarbans of Bangladesh: A multiple evidence base trajectory. In R. Dasgupta, S. Hashimoto, & O. Saito (Eds.), *Assessing, mapping and modelling of mangrove ecosystem services.* Springer Sustainability Science Book Series (Forthcoming).

Titumir, R. A. M., & Paran, M. S. (2022c). Human-nature relationship for post 2020 biodiversity framework: An exploration into traditional knowledge, customary sustainable use and community-based innovation in the

Sundarbans. In R. A. M. Titumir (Ed.), *Fifty years of nation building: Political economy of Bangladesh's development*. Centennial Book Series Project of University of Dhaka. Department of Development Studies, University of Dhaka (Forthcoming).

Titumir, R. A. M., Paran, M. S., & Pasha, M. W. (2019). The Sundarbans is our mind: An exploration into multiple values of nature in conversation with traditional resource users. In UNU-IAS & IGES (Eds.), *Understanding the multiple values associated with sustainable use in socio-ecological production landscapes and seascapes (SEPLS)* (Vol. 5, pp. 97–117). Satoyama Initiative Thematic Review. United Nations University Institute for the Advanced Study of Sustainability.

Unnayan Onneshan. (2020). *Biodiversity, climate change and traditional resource users in the Sundarbans: An exploration through public participation geographic information system (PPGIS)* (Unpublished Research Report). Unnayan Onneshan, Dhaka, Bangladesh.

CHAPTER 2

Current State of Terrestrial Ecosystem: Evidence of Resource Vulnerability

Rashed Al Mahmud Titumir and Tanjila Afrin

Introduction

The Sundarbans has been witnessing extreme ecological and physiographical changes over the past few decades that are taking a heavy toll on its regenerative capacity and sustainability. These changes have caused a continuous decline in forest's density, coverage, and its biodiversity (Islam & Bhuiyan, 2018). This chapter depicts the depletion of resources in the terrestrial ecosystem (shortage of resources available on land) and

R. A. M. Titumir (✉)
Department of Development Studies, University of Dhaka, Dhaka, Bangladesh
e-mail: rt@du.ac.bd; rtitumir@unnayan.org

Unnayan Onneshan, Dhaka, Bangladesh

T. Afrin
Department of Development Studies, Bangladesh University of Professionals, Dhaka, Bangladesh
e-mail: tanjila.afrin@bup.edu.bd

R. A. M. Titumir (ed.), *Sundarbans and its Ecosystem Services*, Sustainable Development Goals Series,
https://doi.org/10.1007/978-981-19-3000-3_2

the driving forces behind it. It is especially true for certain faunal species and wildlife creatures that have gone extinct with time or are on the verge of extinction. The chapter demonstrates that biodiversity of the terrestrial ecosystem of this mangrove forest is rapidly declining and there has been massive degradation of forest coverage and its density in the last two decades.

As it unfolds, the chapter also depicts the growing stock of the mangrove forest according to different inventories and the decadal changes in forest coverage. The study of these aspects helps establish the extent of resource depletion. Based on DPSIR (Drivers, Pressures, States, Impacts, Responses) framework, this chapter defines the cause and effect of biodiversity degradation that can be used to understand and analyse the interdependent relationship between social and environmental factors. It is a well-received popular framework internationally used to explain the present state of a particular ecosystem.

This chapter shows the population density of the forest and regions lying in its proximity such as Satkhira, Khulna and Bagherhat. Besides population study, this chapter also explores the ongoing industrial and development projects. Here, the attempt is to identify as well as analyse the management's approach towards the forest area and the policies that are currently being implemented. This has been done to identify the core issues and challenges and thereby assist in adapting necessary solutions when required. The study incorporated in this chapter has also been done as a means of ensuring sustainable utilisation and conservation of biodiversity resources. These management approaches include the study of factors such as increasing habitation and illegal encroachment, land shortage and land reclamation, rent-seeking tendencies and extra-legal management, marginalisation of native people and existence of poverty, industrialisation and development projects impacting the forest, etc.

The chapter goes on to indicate the presence of convincing evidence of resource vulnerability in the terrestrial ecosystem that has been taken from anthropogenic pressures and natural hazards. Furthermore, this chapter also demonstrates how this forest can be characterised as a Socio-Ecological Production Landscape and Seascape (SEPLS). This has been done by stating a few indicators and identifying whether those indicators are relevant or not and then finally, describing how or why those indicators are relevant. In this case, the indicators are (a) Mosaic of production landscape/seascape and (b) Harmonious interaction between humans and nature and the well-being of both.

Trends in Biodiversity Resources

The world's largest mangrove ecosystem is facing severe a threat of biodiversity loss due to human intervention and natural disorders (Titumir & Paran, 2022a, 2022b, 2022c). Nonetheless, some studies argued that even though the forest's boundary is almost unaffected, its quality is degrading (Hussain & Karim, 1994; Iftekhar & Islam, 2004; Siddiqi, 2001). Between the 1970s and the 2000s, a non-significant reduction (1.1%; 66 km^2 of 6,017 km^2) can be seen in the forestlands of the total Sundarbans based on Landsat satellite data (Aziz & Paul, 2015).

It is obvious that the Sundari (*H. fomes*) tree has declined in both 1989–2000 and 2000–2010 (Table 2.1). The percentage of decline (28.75%) is the maximum amid all other floral species. However, another prominent species of flora, gewa (*Excoecaria agallocha),* increased during that period at quite a small rate (6.7%). According to a tree survey conducted amid 2008–2014, the most abounding mangrove was *E. agallocha* (59.69% of total trees), followed by *H. fomes* (30.89%), *C. decandra* (6.12%) and *X. mekongensis* (0.82%) (Sarker et al., 2016). The main reason behind the decrease of *Sundari* and the increase of gewa and goran plants is the increase in salinity of the forest land. Even though some of the forest species, the marshes and the water catchment areas have increased, the overall forest land coverage has been reduced by 1,598 ha. (3.6%) between 1989 and 2010.

After investigating the vegetation cover of the forest (1975–2006) with the help of Landsat imagery during 1975–1989 and 1989–2000, a recent study by Islam (2014) found that the land area decreased by ~5.1 and ~4.5%, respectively, even though it had increased from 2000 to 2006 by ~3.9%. Another study predicted the mangrove forest coverage (gain and loss) from 2000 to 2012 in South Asia through Landsat satellite data (Giri et al., 2014). This revealed that from 1970 to 2000 the forest coverage had fallen by 1.2%. Also, between 2000 and 2012, the net deforestation decreased, partially due to increased awareness and plantation and forest protection initiatives (Giri et al., 2014). These are two studies that portray a positive picture of the total coverage. In contrast, Islam and Gnauck (2009) reported that in 1776, it had an area of 17,000 km^2 and currently it is almost half of what it was previously.

The decrease in the number of important tree species can also be evaluated through forest inventories (Table 2.2). Three important inventory reports were formulated in 1959, 1983 and 1996 (FAO, 2011). In

Table 2.1 Changes (in ha) of forest cover, marsh and water area from 1989 to 2000 and 2010, determined by GIS technique at Khulna and Chandpai ranges (one spot from each and together)

Classes	*Plant cover and water area (in ha)*			*Changes (in ha)*		
	1989	*2000*	*2010*	*1989–2000*	*2000–2010*	*Overall*
Heritiera fomes	23,028	19,309	16,408	−3,719 (16.15)	−2,901 (15.02)	−6,620 (28.75)
Excoecaria agallocha	15,184	15,828	16,200	644 (4.24)	+372 (02.35)	+1,016 (6.7)
Sonneratia apetala	44	1,906	109	+1862 (4231)	−1,797 (94.28)	+65 (148)
Bruguieras exangula	191	8,169	3,026	+7978 (4177)	−5,143 (63)	+2,835 (14.84)
Marshes	–	387	585	+387 (100)	+198 (33.85)	+585 (100)
Shrubs	570	463	367	−107 (18.77)	−96 (20.73)	−203 (35.61)
Water	4,727	5,581	6,009	+854 (18.07)	+428 (7.67)	+1,282 (27.12)
Total area	44,301	44,328	42,704	+27 (0.06)	−1,624 (3.66)	−1,598 (3.60)

Note Values within parentheses are in per cent
Source Prepared by authors based on Ahmed et al. (2011) and Aziz and Paul (2015)

Table 2.2 Growing stock of the Sundarbans according to different inventories

Year of publication of inventory results	*Inventory done by*	*Sundari (number of trees per hectare)*	*Gewa (number of trees per hectare)*	*All tree species (number of trees per hectare)*
1959	Forest and Forestal Engineering, Canada	211	61	296
1983	Overseas Development Authority	125	35	180
1996	Forest Resource Management Project, FD, GoB	106	20	144

Source FAO (2011)

each case, the growth of trees is declining. The Sundari growing stock has declined by 50%, while Gewa growing stock declined by 67% in 37 years (between 1959 and 1996). Overall, the growing stock of trees has dropped by 51% and the rate of depletion is approximately 1% per year. A recent article (CEGIS, 2016) also revealed the declining tendency of the forest areas between 1776 and 2010, in both India and Bangladesh.

This mangrove forest also provides several types of Non-Wood Forest Products (NWFPs). Table 2.3 reports the production pattern of the main NWFPs over a period.

The production of Golpata increased in 1986–1987 in comparison to 1980–1981 but decreased in 1989–1990 again; and then increased again during 1990–1991 (Table 2.3). A similar trend can be seen in the production of Hental and honey too. Most importantly, the production of fish is

Table 2.3 Production trends of the main NWFPs

Year	*Golpata (in million kg)*	*Hental (in tonnes)*	*Honey (in tonnes)*	*Fish (in tonnes)*	*Shell (in tonnes)*
1980–1981	2.48	6.20	311	–	–
1986–1987	2.63	6.10	229	6.80	1.10
1989–1990	2.48	7.20	147	5.10	1.64
1990–1991	2.63	6.70	211	4.80	1.64

Source Adapted from Roy (2009) and Roy and Alam (2012)

found to be decreasing over time. However, the production of shells was unchanged to some extent.

The deterioration of floral diversity negatively affects the faunal diversity. The forest structure is getting simpler and the mean height of the trees is falling. This leads to a decline in the habitat for birds, monkeys and other such species that mostly live on trees. As many as 20 globally threatened species are here. One of the world's most endangered turtle breeds, *Batagur baska*, is also found in this forest (Behera & Haider, 2012). In addition, there are two other endangered cetacean species, the Ganges river dolphin (*Platanista gangeticus*) and Irrawaddy dolphin. Moreover, other threatened wildlife species include Bengal tiger, python, king cobra, adjutant stork, white-bellied sea eagle, clawless otter, masked fin-foot, ring lizard and river terrapin, fishing cat, spoon-billed sandpiper, eagle and lesser adjutant (Hossain, 2014 as cited in IUCN, 2014).

To add to it, Aziz and Paul (2015) reported that a total of more than 40 species of amphibians, reptiles, avis and mammals are critically endangered or vulnerable. The Royal Bengal Tiger, which is the most important faunal species, is said to be an endangered species by the IUCN. In fact, a limited number of tigers are available in the forest at present and this may further reduce, pushing it towards extinction in the future. Table 2.4 provides a list of animals that went extinct in the last century. From this, it is evident that the forest has been losing its sheen over time, in terms of loss of its extent or coverage and in terms of depletion of its rich biodiversity resources.

Table 2.4 Floral species that became extinct

Scientific name	*Local name*
Rhinoceros sondaicus	Javan rhinoceros
Bubalusbubalis	water buffalo
Cervus duvauceli	swamp deer
Bos gaurus	Guar
Axixporcinus	hog deer
Crocodiles palustric	marsh crocodile

Source Adapted from Behera and Haider (2012)

Decadal Changes in Forest Coverage

The coverage and density of the Sundarbans is declining. The amount of fallow land is increasing due to deforestation. The decadal changes of forest coverage, drawn by PPGIS, indicate that the number of forest trees is declining drastically (Fig. 2.1). As a result, the amount of fallow land is increasing. In two decades, the total area of dense forests in the case study site has halved.

The dark green part of the map refers to the areas of the dense forest, which still have adequate vegetation. The light green part refers to the areas of relatively low or medium dense forests. The amount of vegetation in these areas is less than the dense forest part. On the other hand, the areas where the forest is very thin are marked in white which indicates that the number of trees here is much less or not present at all. The white areas have become empty fallow lands.

The average growth rate of the white part in two decades is about 54.92%. In 2000, the white area in this region was 4,546 hectares. By 2010, the white area had increased by 24.90% to 5,678 hectares. Within a decade, the extent of the white parts has almost doubled in 2020 as compared to 2000. In other words, the amount of fallow land has almost

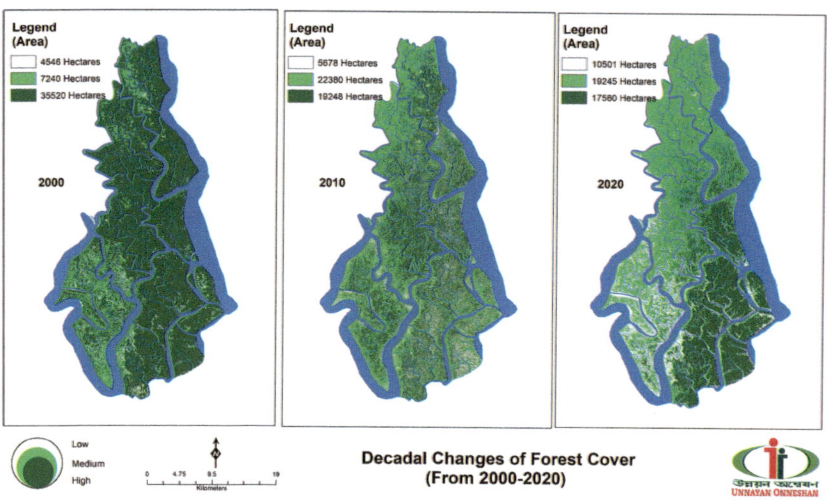

Fig. 2.1 Decadal changes of forest coverage (*Source* Unnayan Onneshan, 2020)

doubled in this region due to rampant deforestation. From 2010 to 2020, the white part has increased by 84.94%. At present, the area of the white part in the case study site is 10,501 hectares (Fig. 2.1).

In contrast, the region of dense green forests has declined sharply in two decades (27.28% on average). In 2000, the amount of dense forest was 35,520 hectares that got reduced by almost half to only 19,248 hectares in 2010. The dark green area of the forest has declined by 45.81% in a decade. The rate of deforestation has declined somewhat over the following decade, but is continuing. From 2010 to 2020, it has decreased by 8.76%. At present, the area of dark green part is 17,560 hectares (Fig. 2.1).

Similarly, there is a marked change in the size of the medium green part. It can be seen that in the first decade (2000–2010) the amount of medium green part has increased by 209.11%. It increased from 7,240 hectares in 2000 to 22,380 hectares in 2010. There has been a slight decline (14%) in the second decade. Now, the area of this part is 19,245 hectares (Fig. 2.1).

On the one hand, the reason for the massive increase in the medium green area in the first decade can be attributed to the decrease in the dark green area at a massive rate (about half), on the other hand, the growth rate of the white part was lower in the first decade than in the second decade. As a result, the dark green part has transformed to medium green. Over the following decade, the white part has almost doubled. The decadal changes of forest coverage ascertain that there is widespread deforestation in this part of Sundarbans. As a result, most of the forest land is becoming empty fallow land.

Issues and Challenges

This section shows the current management approach and critically analyses it in order to identify the main issues and challenges. This is to ensure sustainable utilisation and conservation of biodiversity resources. These issues and challenges are negatively affecting the forest's ecosystem. As a result, vulnerabilities are increasing leading to disruption in the ecosystem. These are mainly from increased human intervention in the forest without considering the forest's sustainability and well-being.

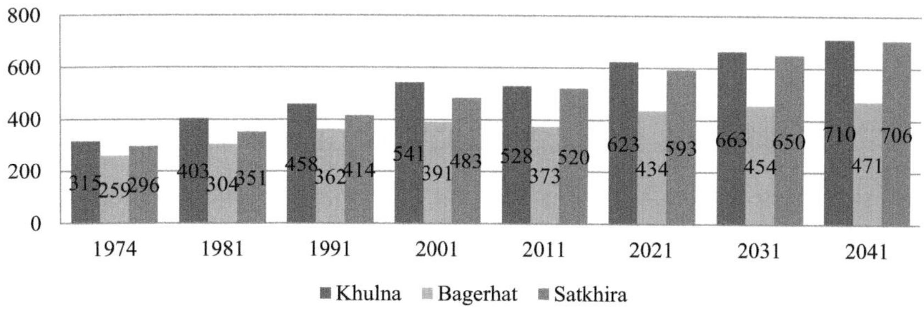

Fig. 2.2 Population density in the encompassing districts of the forest (*Source* Author's calculation based on population census of 2001 and 2011 by BBS, 2001, 2011)

Increasing Habitation and Illegal Encroachment

Due to unstable and ill-defined property rights, the politically and economically powerful groups are intruding illegally into the forest (Titumir et al., 2019, 2022). The density of settlement beyond the Khulna, Bagerhat and Satkhira regions has been rising over the years, and the trend will continue as the projection implies (Fig. 2.2). The upright dependency on natural resources is also increasing. This increasing habitation is mostly the result of weak property rights regime of the community residing in this ecological landscape. A significant number of migrants barge into the forest at regular intervals and settle in the nearby districts.

Since they are not the indigenous local people, they do not know local customary practices of conserving the forest resources. Furthermore, they always attempt to extract as much resource as possible resulting in faster degradation. Other powerful groups continuously force their way into the forest region by affiliating at different levels.

Land Shortage, Land Reclamation and Shrimp Cultivation

The land area of Bangladesh is limited and there is always a struggle to get hold of the land resources. Throughout this process, politically affluent people have better chances of procuring land suitable to them. The bordering regions are probable sites for cultivating shrimps because of the presence of saline water. However, the switching of land into commercial shrimp farms acts as the largest man-made hazard towards the balance of this mangrove ecosystem.

Rent-Seeking Tendencies and Extra-Legal Management

Forest department officials often harass the resource collectors for bribes. Evidence exists to show how timber traders and smugglers are hand in gloves and involved in cutting and selling of trees. In addition to it, poachers with the supposed participation of the forest officials have been openly poaching animals in the region (Titumir et al., 2019).

Marginalisation of Local and Indigenous People and Existence of Poverty

The current management system has marginalised the local and indigenous people in terms of furnishing their rights (Titumir et al., 2020, 2022). The Forest Department is defined by elitist and bureaucratic culture and contains rigorous, hierarchical and top-down working practices. The TRUs can barely enrich their economic conditions, throughout this process. This is because, on one hand, they cannot cultivate resources as per their requirements and on the other hand they must spend money to pay extra tolls, bribes and other illegal claims made by the locally powerful ones (for example, high-interest rate of money lenders, money claimed by dacoits). As a result, the TRUs are compelled to extract resources hampering sustainability for surviving.

Industrialisation and Development Projects Near (or Around) the Forest

In the recent times, powerful entities, both at the national and international levels have successfully coerced the government to approve numerous development projects. Hence, the government has approved the establishment of 190 industrial and commercial units by going against their own policy in the Ecologically Critical Area (ECA) that poses a serious threat to the biodiversity in the area. The government has announced the 10-kilometre periphery of the mangrove forest as an ECA in 1999 after the UNESCO enlisted it to be a natural world heritage site. As per the Bangladesh Environment Conservation Act, 1995 (amended

Table 2.5 Factories near the forest

Types of factory	*Numbers*
Cement factory	6
LPG	7
Gas cylinder	1
Oil refinery	3
Ship building	2
Saw mill	15
Betel nut processing	8
Rice mill	73
Fish farm and hatchery	19
Saline water refinery	7
Brick kilns	3
Others	46

Source *The Daily Star* (2018)

in 2010), nobody is allowed to set up any factories within the ECA (Table 2.5).

Most of these agents and interest groups of land grabbers are businessmen and industrialists with powerful political backing. The most recent and disputed project that is currently being carried out is the 'Rampal Power Plant Project', a coal-based power plant that has been charged for jeopardising all three domains of the environment and in terms of economic and technical feasibility might endanger the bounties of the forest.

Figure 2.3 presents the major challenges or issues that prevail in the current management framework put in place for the conservation of forest resources. Overall, it can be said that a long-term ecological change is happening.

Thus, the 'structure' of the forest can be depicted by demonstrating the key aspects of this mangrove ecosystem which include its location and dynamic nature of various kinds of habitats. The distinct 'benefits' occur in the form of maintaining its biodiversity resources and provisioning of the resources to the Indigenous People and Local Communities (IPLCs) including TRUs (The details of 'structure' and 'benefits' are described in chapter 1). This chapter mainly shows the major 'changes' in the

Table 2.6 The Sundarbans as a SEPLS

Indicators	*Relevant to Sundarbans? (Yes/No)*	*Why relevant?*
Mosaic of production landscape/seascape	Yes	It is a mangrove forest that consists of forest, coastal and wetland ecosystems, supporting diverse production activities
Harmonious interaction between humans and nature and well-being of both	Yes	It provides different choices to the IPLCs to maintain livelihoods and in return, the IPLCs offer protection to the forest and its' resources through traditional livelihood practices

Source Prepared by authors

ecosystem that result from the interactions between humans and nature.[1] Hence, the Sundarbans can be characterised as SEPLS (Table 2.6), as the chapter argues, that has regularly faced threats due to countless forms of pressure.

The overall outline of the forest can also be interpreted in the following diagram under the DPSIR Framework (Fig. 2.4). It demonstrates how all these factors are interlinked with one another, in addition to showing their respective responses. These consist of the driving forces that include industrialisation, constructing flood control embankments (for example, Farakka dam), coal-based power plant for generating electricity, climate change and natural disasters. The pressures include the overexploitation of wood and non-wood products, heavy fishing, extensive use of pesticides, plantation of invasive species, regular oil spillage, illegal encroachment, fire incidents, salinity intrusion, destruction of habitats, shrimp farming, poaching, hunting, etc. The state of the forest talks about the decreasing forest cover extent (though some studies suggest that forest reduction in terms of quantity is non-significant), quality degradation of the ecosystem

[1] The three perspectives drawing on the SEPLS concept—'structure, benefits and changes'—deepen the understanding of various types of production landscape and seascapes (Ichikawa, 2013).

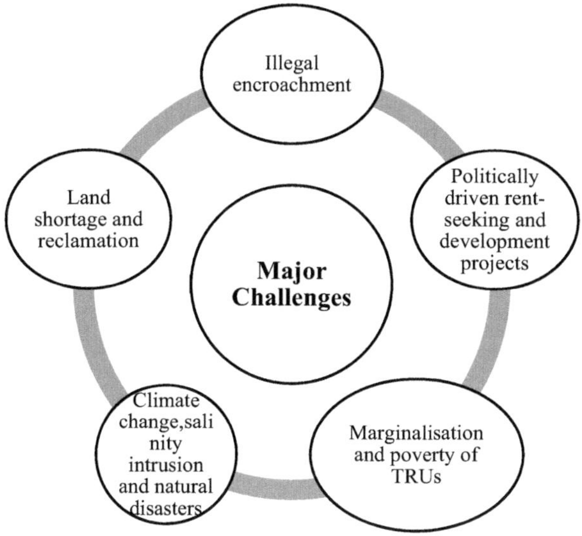

Fig. 2.3 Major challenges in brief (*Source* Prepared by authors)

due to biodiversity loss, some faunal species have already become extinct and some are threatened and reduction of floral species. Finally, the impact talks about the rich and only natural mangrove forest resource of the country that is losing its worth like the loss of its natural beauty, the vulnerable livelihood of local people, loss at the national level in terms of economic and ecological significance. All these factors have been linked with their responses as to how they have been dealt with.

Conclusions

This chapter illustrates that the biodiversity of the terrestrial ecosystem of this mangrove forest is fast depleting. There has been massive degradation of forest coverage in the last two decades, particularly in the Koyra region of the Khulna range. This has been established through empirical evidence which suggests that important tree species and their growth are gradually declining. These include the Sundari and gewa. In addition to this, production trends of the main Non-Wood Forest Products (NWFPs) such as hental, honey, etc., are also deteriorating. All these relate to the deterioration of the flora and fauna and how they are negatively impacted.

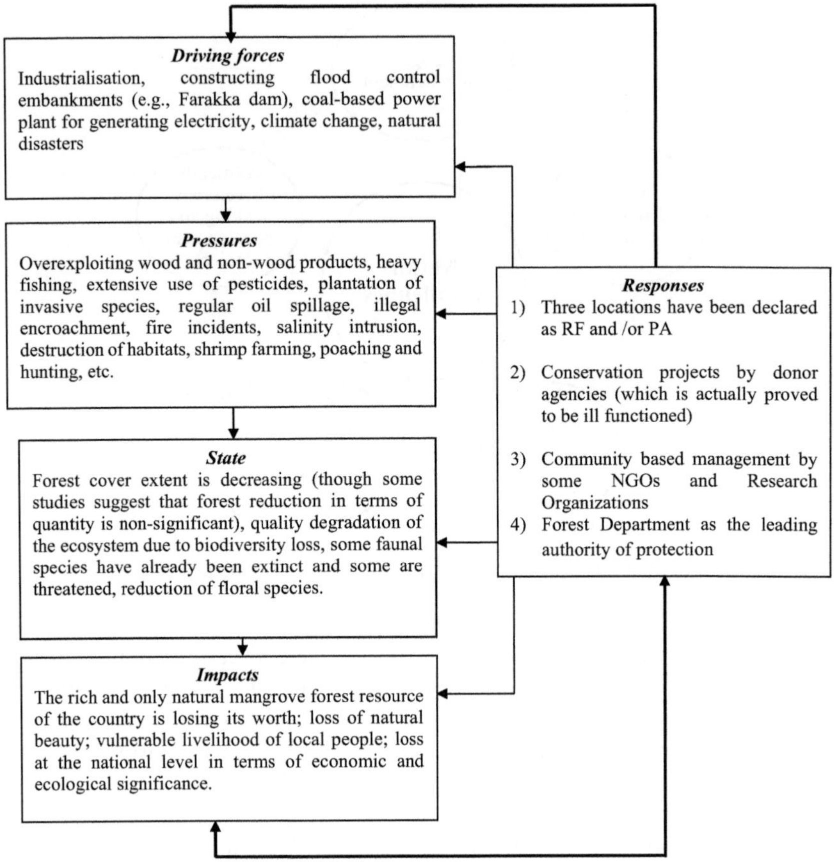

Fig. 2.4 Present state of the forest (*Source* Prepared by authors)

This makes it clear that the Sundarbans has been losing its value over time, in terms of loss of its extent or coverage and of its rich biodiversity resources.

The scrutiny of the present management approach has been defined with complementary challenges and problems. Due to unstable and ill-defined property rights, the politically and economically powerful groups intrude illegally into the forest, for their own benefit. Similarly, due to a shortage of land, it becomes difficult to collect land resources, where the powerful groups also interfere and, in most cases, succeed due to their political influence and monetary competency. Along with these, rent-seeking tendencies, poverty, marginalisation of local people has a cascading impact in disrupting the management approach. It can be demonstrated that in the terrestrial ecosystems of the forest, human activities (anthropogenic pressures) are a cause of resource degradation. Hence, it is crucial to look at the institutional and management system of the forest which overlooks and inquires human activities.

Finally, the Sundarbans can be characterised as a SEPLS that has frequently faced threats due to numerous forms of pressure. Additionally, an outline of the forest is also presented under the DPSIR Framework as that demonstrates its present condition, by stating the factors involved along with their responses.

References

Ahmed, A., Aziz, A., Khan, A. Z. M. N. A., Islam, M. N., Iqubal, K. F., Nazma, M., & Islam, M. S. (2011). Tree diversity as affected by salinity in the Sundarban mangrove forests, Bangladesh. *Bangladesh Journal of Botany, 40*, 197–202.

Aziz, A., & Paul, A. R. (2015). Bangladesh Sundarbans: Present status of the environment and biota. *Diversity, 7*(3), 242–269.

Bangladesh Bureau of Statistics. (2001). *Population and housing census 2001.* Statistics and Informatics Division (SID), Ministry of Planning, Government of the People's Republic of Bangladesh, Dhaka, Bangladesh.

Bangladesh Bureau of Statistics. (2011). *Population and housing census 2011.* Statistics and Informatics Division (SID), Ministry of Planning, Government of the People's Republic of Bangladesh, Dhaka, Bangladesh.

Behera, M. D., & Haider, M. S. (2012). *Situation analysis on biodiversity conservation—Ecosystem for life—A Bangladesh India initiative.* IUCN.

CEGIS. (2016). *Joint landscape narrative by India and Bangladesh.* Center for Environmental and Geographic Information Services (CEGIS), Unpublished Report, Dhaka, Bangladesh.

FAO. (2011). *Bangladesh forestry outlook study: Asia-Pacific Forestry Sector Outlook II* (p. 101) (Working Paper Series, Paper No. APFSOS II/WP/2011/33).

Giri, C., Long, J., Abbas, S., Murali, R. M., Qamer, F. M., Pengra, B., & Thau, D. (2014). Distribution and dynamics of mangrove forests of South Asia. *Journal of Environmental Management, 148*, 101–111.

Hussain, Z., & Karim, A. (1994). Introduction. In Z. Hussain & G. Acharya (Eds.), *Mangroves of the Sundarbans* (Vol. 2: Bangladesh). IUCN.

Hussain, M. Z. (2014). *Bangladesh Sundarban delta vision 2050: A first step in its formulation—Document 2: A compilation of background information*. International Union for Conservation of Nature (IUCN), Bangladesh Country Office, Dhaka, Bangladesh.

Ichikawa, K. (2013). Understanding socio-ecological production landscapes in the context of Cambodia. *International Journal of Environmental and Rural Development, 4*(1), 57–62.

Iftekhar, M. S., & Islam, M. R. (2004). Degeneration of Bangladesh's Sundarbans mangroves: A management issue. *The International Forestry Review, 6*(2), 123–135.

Islam, M. S. N., & Gnauck, A. (2009). Threats to the Sundarbans mangrove wetland ecosystems from transboundary water allocation in the Ganges Basin: A preliminary problem analysis. *International Journal of Ecological Economics and Statistics, 13*(9), 64–78.

Islam, M. T. (2014). Vegetation changes of Shundarbans based on landsat imagery analysis between 1975 and 2006. *Landscape & Environment, 8*(1), 1–9.

Islam, S. M. D., & Bhuiyan, M. A. H. (2018). Sundarbans mangrove forest of Bangladesh: Causes of degradation and sustainable management options. *Society for Environmental Sustainability, 1*, 113–131.

IUCN. (2014). *Bangladesh Sundarban delta vision 2050: A first step in its formulation*. Bangladesh Country Office.

Roy, A. K. D. (2009). *Wetland management and valuation: The Sundarbans perspective for participatory forestry*. Academic Press and Publishers Library.

Roy, A. K. D., & Alam, K. (2012). Participatory forest management for the sustainable management of the Sundarbans mangrove forest. *American Journal of Environmental Science, 8*(5), 549–555.

Sarker, S. K., Reeve, R., Thompson, J., Paul, N.K., & Matthiopoulos, J. (2016). *Are we failing to protect threatened mangroves in the Sundarbans world heritage ecosystem?* (Scientific Reports).

Siddiqi, N. A. (2001). *Mangrove forestry in Bangladesh*. Institute of Forestry and Environmental Sciences, University of Chittagong, Bangladesh.

The Daily Star. (2018). Licence to harm Sundarbans. https://www.thedailystar.net/frontpage/licence-harm-sundarbans-1558918. Accessed on 23 March 2021.

Titumir, R. A. M., Afrin, T., & Islam, M. S. (2020). Traditional knowledge, institutions and human sociality in sustainable use and conservation of biodiversity of the Sundarbans of Bangladesh. In O. Saito, S.M. Subramanian, H. Hashimoto & K. Takeuchi (Eds.), *Managing socio-ecological production landscapes and seascapes for sustainable communities in Asia: Mapping and navigating stakeholders, policy and action* (pp. 67–92). Science for Sustainable Societies. https://doi.org/10.1007/978-981-15-1133-2_5

Titumir, R. A. M., Afrin, T., & Islam, M. S. (2022). *Natural resource degradation and human-nature wellbeing: Cases of biodiversity resources, water resources, and climate change*. Springer (Forthcoming).

Titumir, R. A. M., & Paran, M. S. (2022a). Human-nature cooperation for well-being: Community understanding on one health approach in the COVID-19 era in the Sundarbans. In M. Nishi, S. M. Subramanian, H. Gupta (Eds.), *Biodiversity-health-sustainability nexus in socio-ecological production landscapes and seascapes (SEPLS)*. Springer Nature.

Titumir, R. A. M., & Paran, M. S. (2022b). Ecosystem services and well-being in the Sundarbans of Bangladesh: A multiple evidence base trajectory. In R. Dasgupta, S. Hashimoto, & O. Saito (Eds.), *Assessing, mapping and modelling of mangrove ecosystem services*. Springer Sustainability Science Book Series (Forthcoming).

Titumir, R. A. M., & Paran, M. S. (2022c). Human-nature relationship for post 2020 biodiversity framework: An exploration into traditional knowledge, customary sustainable use and community-based innovation in the Sundarbans. In R. A. M. Titumir (Eds.), *Fifty years of nation building: Political economy of Bangladesh's development*. Centennial Book Series Project of University of Dhaka. Department of Development Studies, University of Dhaka (Forthcoming).

Titumir, R. A. M., Paran, M. S., & Pasha, M. W. (2019). The Sundarbans is our mind: An exploration into multiple values of nature in conversation with traditional resource users. In UNU-IAS & IGES (Eds.), *Understanding the multiple values associated with sustainable use in socio-ecological production landscapes and seascapes (SEPLS)* (Vol. 5, pp. 97–117). Satoyama Initiative Thematic Review. United Nations University Institute for the Advanced Study of Sustainability.

Unnayan Onneshan. (2020). *Biodiversity, climate change and traditional resource users in the Sundarbans: An exploration through public participation geographic information system (PPGIS)* (Unpublished Research Report). Unnayan Onneshan, Dhaka, Bangladesh.

CHAPTER 3

Current State of Biodiversity in Marine and Coastal Ecosystem of the Sundarbans

Rashed Al Mahmud Titumir and Md. Shah Paran

Introduction

This chapter aims to illustrate the contemporary state of biodiversity in the marine and coastal ecosystems in the Sundarbans. It attempts to take into account the perspective of the Traditional Resource Users (TRUs) in the forest to depict the current state of aquatic resources at coastal and marine ecosystem. The indigenous and local knowledge on the state of aquatic resources has been collected through participant observation, key person interviews and focused group discussions. While curating the content of the chapter, scientific evidence from secondary sources has

R. A. M. Titumir (✉)
Department of Development Studies, University of Dhaka, Dhaka, Bangladesh
e-mail: rt@du.ac.bd; rtitumir@unnayan.org

R. A. M. Titumir · Md. S. Paran
Unnayan Onneshan, Dhaka, Bangladesh

R. A. M. Titumir (ed.), *Sundarbans and its Ecosystem Services*, Sustainable Development Goals Series,
https://doi.org/10.1007/978-981-19-3000-3_3

been used as well. In addition, PPGIS has been adopted for the ecological value mapping of fish and crabs.

There has always been a lack in incorporating the perspective of the TRUs in understanding the biodiversity of this mangrove forest. With the help of examination of the present scenario of coastal and marine biodiversity, this study helps to raise the issue of aquatic resource degradation. The chapter reveals that in the last two decades (2000–2020), the amount of crabs and big fish in the Sundarbans has decreased. The degradation of aquatic resources is mainly due to the over-extraction of resources through unsustainable means accelerated by profit making impulses of people. Both aquatic floral and faunal species have declined significantly in terms of quality and quantity.

The chapter reveals that significant damage is being inflicted on the ecosystem due to multiple anthropogenic pressures. A major contributor towards inflicting such damage is the rent-seeking powerful syndicate comprising numerous administrative and political agencies exploiting the resources. It is further demonstrated how such behaviour, if not stalled and if it continues to prevail, will make it impossible to conserve the valuable resources of the forest, as well as the biodiversity and ecosystem.

At the beginning of the chapter, the current state of biodiversity has been discussed with the help of scientific evidence. A list comprising important aquatic species has been made and later it is described how and why they have declined because of factors such as anthropogenic pressures or over-exploitation. It goes on to put on record the ecological value of different parts of the water bodies based on the availability of aquatic resources such as big fish and crabs to show how the value of different parts has been declining and resources are depleting in the past two decades (2000–2020). Later in the chapter, the TRUs' perspective has been highlighted in regard to community-based monitoring and management of both marine and coastal biodiversity of the ecosystem. It shows how the TRUs are dependent on the forest for their livelihood and how they collect resources from the forest using traditional knowledge and their own methods passed on through generations. In addition to these, the TRUs' opinion on the present condition of the coastal and marine ecosystem has been presented in this chapter providing an insight into their way of thinking which has helped them to live in harmony with the forest. Moreover, based on the opinion and viewpoint of the TRUs, a DPSIR framework depicting a cause-and-effect relationship has been formulated in regard to the loss of coastal and marine biodiversity.

This framework includes different parameters such as the Drivers, Pressure, State, Impact and Response (DPSIR) in the forest to portray how they are interlinked with one another.

Current State of Biodiversity: Scientific Evidence

Due to anthropogenic pressures coupled with fragile institutions and an ineffective command-driven governance system, the majority of the species in the forest have declined (Titumir & Afrin, 2017; Titumir et al., 2019, 2022). The IUCN list regarding major fish species along with their current state portrays a threatening image (Table 3.1). Almost all the major fish species are susceptible to be endangered. This loss of biodiversity is caused by excessive human interference. Unfortunately, the situation is deteriorating over time due to the incessant extraction of aquatic resources through unsustainable means. The total annual fisheries landing, along with the offshore catch is estimated to be around 75,000MT (Hoq, 2007). Such high amounts of fish extraction are proving to be hazardous to the entire coastal and marine ecosystems of the forest. Table 3.1 shows the threatened fish species, categorised

Table 3.1 Threatened fish species of the Sundarbans (modified from IUCN)

Family	*Species*	*Status*	*Global status*
Ambassidae	*Pseudam bassisbaculis*	VU	–
	P. ranga	VU	–
Anguillidae	*Anguilla bengalensis*	VU	–
Carcharhindiae	*Carcharhinus limbatus*	–	VU
Glyphisganeticus		–	CR
Eleotridae	*Butis butis*	NO	LR
Plotosidae	*Plotosus canius*	VU	–
Pristidae	*Pristis microdon*	–	EN
Scatophagidae	*Scatophagusargus*	EN	–
Schilbeidae	*Pangasius pangasius*	CR	–
Siloniasilondia		EN	–
Syngnathidae	*Hippocampus kuda*	–	VU
Microphisdeocata		EN	–

Source Adapted from Hoq (2007)
Note CR = critically endangered; EN = endangered; LR = lower risk; NO = not threatened; VU = vulnerable

according to their family and in terms of their statuses which help to understand whether they are vulnerable, threatened, critically threatened, endangered, not threatened, at lower risk, etc.

Most of the major available fish species in the region have been classified as overexploited, especially those with a higher market price. A few species have already been exploited profusely, while some are in the optimum level of extraction. The estimation of yield, rate of exploitation and MSY (Maximum Sustainable Yield) of major shell and finfish species in SRF (Sundarbans Reserve Forest) suggests that there is a serious deterioration in aquatic biodiversity of the coastal and marine ecosystems (Table 3.2). The estimation of yield, exploitation rate and MSY of major shell and finfish species in SRF is given in Table 3.2. It portrays different species and represents each parameter for them accordingly.

Table 3.2 Estimation of yield, exploitation rate and MSY of major shell- and finfish species in SRF

Species	*Yield (tons)*	*Exploitation rate*	*MSY (tons)*	*Remarks*
Tenualosa ilisha	762	0.41	523	Over exploited
Lates calcarifer	150	0.35	160	Fully exploited
Pomadasys hasta	232	0.40	457	Optimum
Johnius argentus	548	0.47	593	Optimum
Pangasius pangasius	135	0.42	92	Over exploited
Plotossus canius	141	0.36	92	Over exploited
M. rosenbergii	274	0.30	711	Optimum
Penaeus monodon	180	–	226	Optimum
Scylla serrata	375	–	283	Over exploited
Oyster	3000	–	6000	Under exploited
Gastropod	35	–	113	Under exploited
P. monodon fry	1453 millions	–	672 millions	Over exploited

Source Adapted from Hoq (2007)

Therefore, it could be asserted that major shell and finfish species are either overexploited or face the threat of being at a critical level of extraction, leading to severe biodiversity degradation. As fish and other aquatic resources of the forest are in high demand, it induces pressure on the ecosystem. Limited natural resource coupled with high demand results in skyrocketing market price which in turn triggers over exploitation of resources completing the vicious cycle (Titumir et al., 2019). The high commercial value of these resources generates income for the people who are extracting resources, overseeing the significant loss in biodiversity. The massive share of this profit is being used to grease the palms of a powerful class or group consisting of administrative and political agencies but in the meantime, traditional resource users are being left deprived.

In addition to anthropogenic pressures, natural hazards also pose big risks to this forest (Titumir & Paran, 2022a, 2022b, 2022c). Climate change is severely affecting the ecosystem by negatively impacting the resources and triggering biodiversity loss. Salinity is increasing at a fast pace as the sea water level is rising and there is a low flow of freshwater from upstream rivers due to the construction of embankments (Aziz & Paul, 2015; World Bank, 2016). The water flow of the Ganges reduced from 3,700 m^3/s in 1962 to only 364 m^3/s in 2006 (Islam & Gnauck, 2009). These interventions resulting in higher salinity have affected the habitat of different species that are negatively impacting the ecosystem. The rate at which natural disasters occur has also increased significantly over time, with cyclones frequently hitting the coastal belt of Bangladesh which is leading to degradation of the mangrove ecosystem. In the last 135 years, more than 45 cyclones have hit the coastal belt of Bangladesh (IUCN, 2014). Such hazards ruin the ecosystem of the forest, thus ruining its biodiversity.

Ecological Value Mapping of Fishing

According to the fishermen in the case study area, now big fish are seldom found in the rivers, canals or creeks throughout the forest. These fish are mostly confined to the areas where the colour of the water is pitch dark or muddy and the flow of water is high. The areas where big fish are found are more valuable to the fishermen. The dark red part in the figure (3.1) depicts the most resource-rich areas of rivers, ditches, canals and creek along the forest. The medium red part indicates the areas where big fish could be found but lesser than the dark red parts. Some big fish can also be seen in the areas marked in light red. The larger the red area in the figure, the greater the number of big fish caught in those areas (Fig. 3.1).

Ecological value mapping of big fish suggests that the amount of big fish has declined in the last two decades (2000–2020). According to the information given by the fishermen during PPGIS and the mapping done based on this data ascertains the fact that in the past two decades (2000–2020) the value of aquatic bodies of the forest, consisting of rivers, canals and creeks has decreased over time based on the availability of prized fish catch like the big fish (Fig. 3.1). As the quantity of big fish has decreased in different regions, the value of the area has also dwindled.

In the past two decades, the average reduction rate of areas marked with the dark red colour was 20.98%. In 2000, the area under the dark red part was 520 hectares. Within ten years, this area has decreased by 13.84% to 448 hectares in 2010. In the following ten years, it decreased by 28.12% and stood at 322 hectares in 2020.

Similarly, in the past two decades, the size of medium and light red parts has decreased on an average of 22.18 and 15.09%, respectively. In 2000, the medium red areas were 1,521 hectares. In a decade, it decreased by 19.19% and in 2010 the area stood at around 229

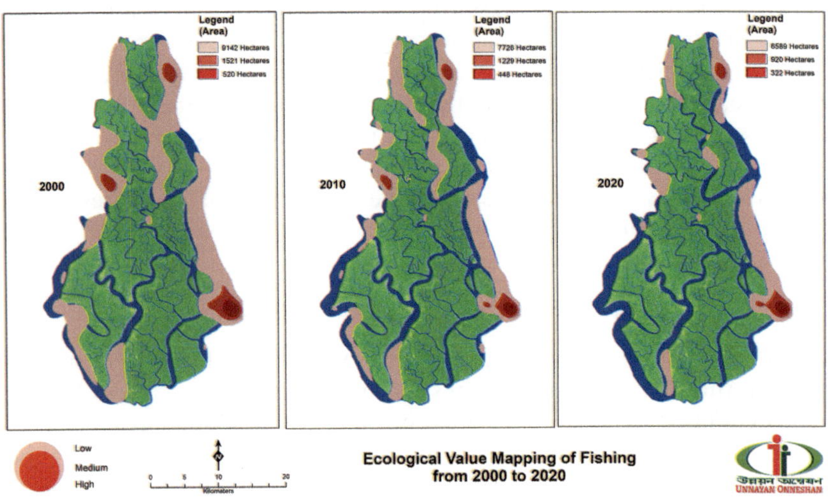

Fig. 3.1 Distribution of fish (ecological value mapping of fishing) (*Source* Unnayan Onneshan [2020])

hectares. In the period from 2010–2020, the medium red part has further decreased to 920 hectares (25.14%). On the other hand, the size of the light red part in 2000 was 9,142 hectares that has decreased by 15.48% to 628 hectares in 2010. In the following decade, there was a further reduction of 14.7% restricting it to 6,589 hectares in 2020 (Fig. 3.1).

The above data shows that the different portions of the water bodies in terms of value attached to big fish have been declining gradually for two decades. The declining value of various parts implies that the number of big fish in the forest has decreased over the years.

Ecological Value Mapping of Crabs

Crabs are not found in all rivers, canals or creeks. They are more common in areas where the colour of the water is dark black or muddy. Areas, where crabs are found in abundance, are more valuable to the crab collectors, as more crabs are extracted from these areas. The dark red part in the Fig. 3.2 points out the most valuable part where crabs are found in abundance. The medium red part means that crabs are also found here but less than the dark red part. Crabs are also found in the light red portions but less than medium and dark parts. The higher the amount of red part, the greater the amount of crab is harvested from those areas (Fig. 3.3).

Ecological value of crabs suggests that the amount of crabs in different waterbodies of the Sundarbans has declined in the last two decades (2000–2020). According to the data collected from crab collectors, the value of various rivers, canals, creeks, etc., of the forest to the crab collectors witnessed a slight increase in the first decade (2000–2010) based on the availability of crabs. However, in the second decade, the value of such areas has decreased a lot. In other words, in the first decade, the quantity of crabs was more in the waterbodies but the same decreased in the second decade. As the quantity of crabs has decreased, the value of these areas has also gone down (Fig. 3.2).

In 2000, the area under dark red part was 2,948 hectares. In 2010, this area witnessed an increase by 6.68% to 3,145 hectares. But in the following decade (2010–2020) the size of dark red part dropped significantly and almost halved (44.86%), recording a decrease from 3,145 hectares in 2010 to only 1,634 hectares in 2020 (Fig. 3.2).

The medium red part has declined at an average of 14.88% in two decades. In 2000, this area was around 13,220 hectares. In a decade, it decreased by 8.32% and in 2010 it stood at around 12,120 hectares.

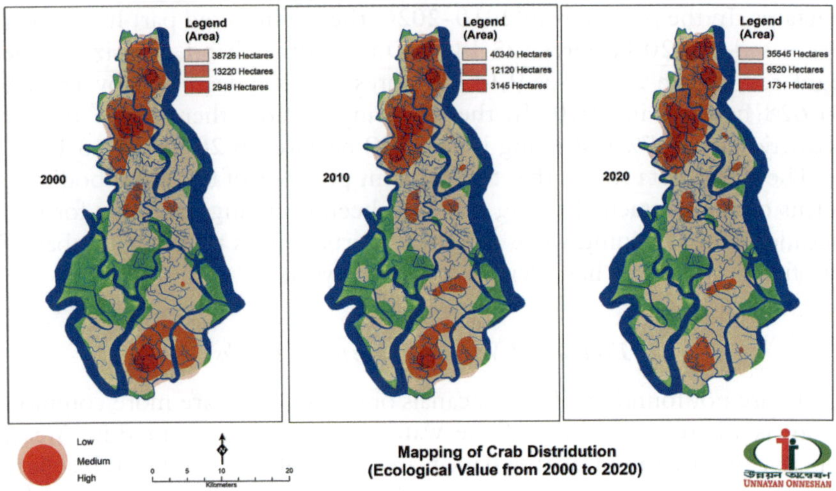

Fig. 3.2 Distribution of crabs (ecological value mapping) (*Source* Unnayan Onneshan [2020])

In the following decade, the medium red part has further decreased by 21.45% standing at around 9,520 hectares in 2020. Meanwhile, the amount of light red part has increased slightly in the first decade (4.16%). But in the second decade, it decreased again by 11.88% and stood at around 35,545 hectares in 2020.

The data given implies that value of waterbodies in the forest has been declining in terms of availability of crabs. This is due to incessant crabbing in areas with a better catch that gradually turns the more valuable areas into less valuable ones.

Traditional Resource Users' Perspective: Community-Based Monitoring

The present condition of biodiversity loss in coastal and marine ecosystem is also evident from the opinion of the TRUs. They pointed out the comprehensive scenario of coastal and marine ecosystem, both causes and effect, inclusively in their own words. TRUs rely on the forest in different ways (see Titumir & Paran, 2022a, 2022b, 2022c for more). They get variety of fish, shrimp, prawn, crabs, shells and other aquatic resources from the forest. Focus group discussions and interviews revealed that as the resource declined significantly, the amount of collection by the TRUs has also gone down. TRUs also found that some common species of fish that were present in the past are not available as catch at present. Therefore, the current state of the resource reveals the adverse scenario of the biodiversity existing in the coastal and marine ecosystems. The TRUs' opinions on the present state of the coastal and marine ecosystem is summarised in Table 3.3.

TRUs' assessments propose that climate change is affecting the forest significantly (see chapter 8 for more). Salinity is rising, while the flow of fresh water is falling. Water bodies are constantly getting polluted. All these factors in unison are contributing to biodiversity loss, in terms of both quality and quantity. The increase in salinity has been a big factor of biodiversity loss in the coastal and marine ecosystems. This also influences the distribution of biodiversity to a big extent. The change in climate patterns and the increase in the number of natural disasters are also affecting the coastal belt. The magnitude of anthropogenic pressures and

Table 3.3 TRUs' opinion on the current state of coastal and marine ecosystem

TRUs opinion
"If there is more flow of water in the rivers, the rivers become filled with fish. Hence, more fish can be caught. As streams of water have been declining, the number of fish and other aquatic resource collection from the forest has also declined. Powerful groups are making embankments that are reducing the water flow."
"If the water in the rivers is muddy, then the number of fish and other aquatic resources increases. On the contrary, the colour of the water is becoming black day by day; this hampers the breeding and habitation of aquatic flora and fauna."
"Fishermen are highly paid if they supply larvae and carps to nearby commercial fisheries. Excess demand by the fisheries causes more extraction of larvae and carps of different fish. If this goes on, there will be no fish left in the rivers."
"Powerful groups, most of them coming from outside areas, are using '*Bishali Jaal*', '*Bain Jaal*', '*Net Jaal*' and '*Chor Jaal*' for fishing. These fishing nets destroy larvae, carps and other small aquatic fauna. Although these harmful nets are prohibited, it is still being used by bribing the forest department."
"These rent-seeking powerful groups poison the water to catch more fish and aquatic resources at a time destroying the aquatic species. This poisoning malpractice ruins the development and habitations of aquatic species."
"After the cyclone Aila, the salinity of this area has increased significantly. Before that, people could use freshwater from ponds. However, even pond water has become saline after the cyclone Aila."

Source Compiled by authors

natural hazards is also increasing. This leads to the massive degradation of biodiversity and the ecosystem.

As per the TRUs, the resources found in the region attract great monetary value. Therefore, people in the area are exploiting the resources to the brim in order to earn more. This marketisation is giving rise to rampant human intervention in the forest, by people entering the region from outside areas. Such people turn a blind eye towards conservation and sustainability of the available resources. Khalilur Rahman, one of the respondents said,

> Selling resources from the forest is becoming profitable progressively. On the other hand, the amount of resources is declining. People are extracting resources as much as they can so that they can earn more profit. They are using various illegal and unfair means for this.

The respondents also said that their collection processes are not harmful towards the conservation of the forest (see Chapters 10.1007/978-981-19-3000-3_4 and 10.1007/978-981-19-3000-3_5). Fatema Begum, a female respondent said,

> We consider forest as our kith and kin. So causing harm to the forest is like causing harm to us. We collect the resources from the forest in such a way that does not harm the forest, rather helps to sustain.

The laws, rules and regulations that are in place concerning human access into the forest have restricted the movements of TRUs, who are dependent on these resources since antiquity. Meanwhile, powerful groups are exploiting these resources illegally through bribing administrative bodies responsible for guarding the resources from exploitation. Omar Ali added,

> Those who are powerful either politically or financially are bribing the forest department. They are extracting resources illegally. They do not care about the survival of the forest. Forest department is blind to them.

This rent-seeking group remains beyond laws and regulations. Moreover, they can significantly cause an influence in the decision-making process of forest management while TRUs' views were ignored. Anju Ara, another female TRU said,

Fig. 3.3 A canal in the Sundarbans (*Photo* Unnayan Onneshan)

> They are all. They are doing whatever they want, what is profitable to them. We, TRUs, are the victim. No one listens to us.

Loss of Coastal and Marine Biodiversity: The DPSIR Framework

Based on the TRUs' opinions and secondary evidences, the biodiversity loss in the coastal and marine ecosystem of the Sundarbans can be illustrated completely with the cause-and-effect relationship by a DPSIR framework. Specific driving forces induce pressure on the ecosystem that creates a particular situation. The driving forces are mostly outcomes of human activities resulting from profiteering motives. Rent seeking behaviour of the powerful class captivates them to extract resources, without thinking about the conservation of the ecosystem. The responses taken as solutions to these problems appear to be faulty. In most cases, these responses are adversely affecting the forest. Figure 3.4 portrays the present condition of coastal and marine ecosystems.

Conclusions

The chapter opens by depicting the current state of biodiversity which is backed by scientific evidence. From the chapter, it is easily concluded as

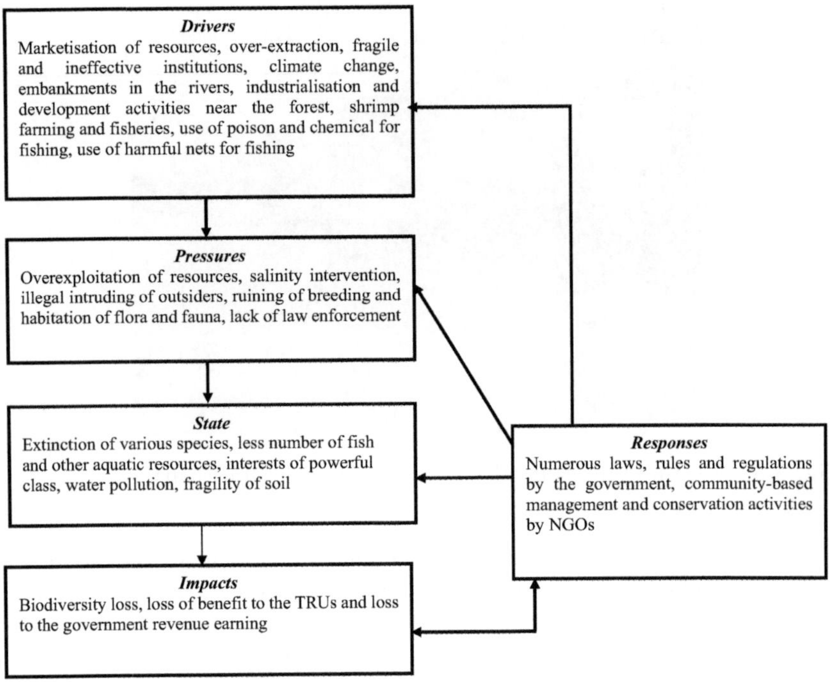

Fig. 3.4 DPSIR framework for coastal and marine ecosystem (*Source* Prepared by authors)

to how valuable resources are declining in the region given to incessant extraction of aquatic resources. The biodiversity loss in the marine and coastal ecosystems of the Sundarbans is the outcome of illegal and unsustainable human activities. This implies that if proper action is not taken regarding the poor state of affairs within the forest then it will be difficult to conserve these valuable resources and in turn protection of this mangrove forest will be jeopardised in the days to come. Such exploitative behaviour will hamper the ecosystem and its biodiversity in numerous ways.

The biodiversity loss is in terms of loss of various species which have been overexploited threatening their existence. Both floral and faunal aquatic species are facing significant decline in terms of quality and quantity. Selfish profiteering motives of people are forcing them to extract the resources in an unsustainable manner. Anthropogenic pressure coupled with climate change is ruining the habitat and breeding conditions of many species in the coastal and marine regions as a result of which its resources are depleting.

The chapter also incorporates the opinions and perspectives of the TRUs, who are dependent on the forest and are concerned about it. They have shown a comprehensive scenario of coastal and marine ecosystems, both causes and effect, inclusively with the help of their own understanding. Their opinions included a range of topics from how the rivers could be used effectively to how influential groups are hampering the biodiversity through harmful practices and unwanted resource exploitation. They also talked about how climate change has affected the forest adversely. In addition, the TRUs also stated how marketisation and the drive to earn more profits have given rise to the number of outsiders entering the forest and overexploiting its valuable resources through improper means. They also added how causing harm to the forest meant causing harm to their own lives that reflect the importance and value they attach to the mangrove forest. These words and feedback of the TRUs help to explain the extent of unfairness and exploitation that may have been happening in the forest for quite some time.

Lastly, this chapter also used a DPSIR framework which has depicted the overall state with the help of a cause-and-effect relationship for the coastal and marine ecosystems in the forest to reveal the loss of biodiversity. This framework has been based on the TRUs' viewpoints and secondary evidences. Moreover, this includes the driving forces, impact, etc., along with the combined responses which are also considered as the solutions, even though most appear to be faulty.

References

Aziz, A., & Paul, A. R. (2015). Bangladesh Sundarbans: Present status of the environment and Biota. *Diversity, 7*(3), 242–269.

Hoq, M. E. (2007). An analysis of fisheries exploitation and management practices in Sundarbans mangrove ecosystem, Bangladesh. *Ocean and Coastal Management, 50*, 411–427.

Islam, M. S. N., & Gnauck, A. (2009). Threats to the Sundarbans mangrove wetland ecosystems from transboundary water allocation in the Ganges basin: A preliminary problem analysis. *International Journal of Ecological Economics and Statistics, 13*(9), 64–78.

IUCN. (2014). *Bangladesh Sundarban delta vision 2050: A first step in its formulation*. Bangladesh Country Office. Dhaka, Bangladesh.

Onneshan, U. (2020). *Biodiversity, climate change and traditional resource users in the Sundarbans: An exploration through public participation geographic information system (PPGIS)* (Unpublished Research Report).

Titumir, R. A. M., & Afrin, T. (2017). Complementarities of human-nature well-beings: A case illustrated through traditional forest resource users of sundarbans in Bangladesh. *Satoyama Initiative Thematic Review (SITR)—Volume 3*, United Nations University Institute for the Advanced Study of Sustainability (UNU-IAS): Tokyo, Japan.

Titumir, R. A. M., Afrin, T., & Islam, M. S. (2020). Traditional knowledge, institutions and human sociality in sustainable use and conservation of biodiversity of the Sundarbans of Bangladesh. In O. Saito, S.M. Subramanian, H. Hashimoto & K. Takeuchi (Eds.), *Managing Socio-ecological Production Landscapes and Seascapes for Sustainable Communities in Asia: Mapping and Navigating Stakeholders, Policy and Action* (pp. 67–92). Science for Sustainable Societies. https://doi.org/10.1007/978-981-15-1133-2_5

Titumir, R. A. M., Afrin, T., & Islam, M. S. (2022). *Natural resource degradation and human-nature wellbeing: Cases of biodiversity resources, water resources, and climate change*. Springer (Forthcoming).

Titumir, R. A. M., & Paran, M. S. (2022a). Human-nature cooperation for well-being: Community understanding on one health approach in the COVID-19 era in the Sundarbans. In M. Nishi, S. M. Subramanian, H. Gupta (Eds.), *Biodiversity-health-sustainability nexus in socio-ecological production landscapes and seascapes (SEPLS)*. Springer Nature, Singapore.

Titumir, R. A. M. & Paran, M. S. (2022b), 'Ecosystem services and well-being in the Sundarbans of Bangladesh: A multiple evidence base trajectory. In R. Dasgupta, S. Hashimoto & O. Saito (Eds.), *Assessing, mapping and modelling of mangrove ecosystem services*. Springer Sustainability Science Book Series (Forthcoming).

Titumir, R. A. M., & Paran, M. S. (2022c). Human-nature relationship for post 2020 biodiversity framework: An exploration into traditional knowledge, customary sustainable use and community-based innovation in the Sundarbans. In R. A. M. Titumir (Ed.), *Fifty years of nation building: Political economy of Bangladesh's development*. Department of Development Studies. Centennial Book Series Project of University of Dhaka, University of Dhaka (Forthcoming).

Titumir, R. A. M., Paran, M. S., & Pasha, M.,W. (2019). The Sundarbans is our mind: an exploration into multiple values of nature in conversation with traditional resource users. In UNU-IAS & IGES (Eds.), *Understanding the multiple values associated with sustainable use in socio-ecological production landscapes and seascapes (SEPLS)* (Vol. 5, pp. 97–117). Satoyama Initiative Thematic Review. United Nations University Institute for the Advanced Study of Sustainability, Tokyo, Japan.

World Bank. (2016). *Impact of climate change and aquatic salinization on mangrove species and poor communities in the Bangladesh Sundarbans* (World Bank Policy Research Paper 7736).

CHAPTER 4

Traditional Knowledge and Customary Sustainable Practices in Terrestrial Ecosystem

Rashed Al Mahmud Titumir and Tanjila Afrin

Introduction

This chapter explores the nuances of traditional knowledge and sustainable practices that have been in place since ages to conserve the terrestrial resources of the forest. A harmonious cohabitation between humans and nature can be assured by letting the local and indigenous people apply their Indigenous Local Knowledge (ILK) in sustainably utilising the forest resources. It goes without saying that traditional knowledge of the locals

R. A. M. Titumir (✉)
Department of Development Studies, University of Dhaka, Dhaka, Bangladesh
e-mail: rt@du.ac.bd; rtitumir@unnayan.org

Unnayan Onneshan, Dhaka, Bangladesh

T. Afrin
Department of Development Studies, Bangladesh University of Professionals, Dhaka, Bangladesh
e-mail: tanjila.afrin@bup.edu.bd

R. A. M. Titumir (ed.), *Sundarbans and its Ecosystem Services*, Sustainable Development Goals Series,
https://doi.org/10.1007/978-981-19-3000-3_4

has been acquired through experience formulated over ages by living in close contact with the natural environment. Hence, it can be said that the locals will know and deploy the best sustainable strategies while using the forest resources and the interdependence they share with the forest suggests that they would continue doing so in the future, ensuring the safety of the forest, by stalling overexploitation and misuse of its resources. Incorporation of ILK for conservation might turn out to be a good initiative as this will include people linked to the forest, who possess better knowledge and are aware of the forest and its resources.

The chapter begins by portraying a few traditional practices followed by the TRUs at the time of harvest. TRUs follow certain rules, passed through the generation of traditional practices while harvesting forest resources. They have certain traditional customs and beliefs which also emphasise resource conservation. In addition to it, the TRUs involve themselves in innovating newer techniques and processes for collecting forest resources using their traditional knowledge. These innovations are being brought in to develop new harvesting techniques and tackle challenges like climate change and fast-evolving geographical conditions in the region. Against this backdrop, the discussions in the chapter are divided into the following three segments: (a) traditional practices employed by TRUs to harvest resources, (b) cultural beliefs of the TRUs and IPLCs and (c) innovation introduced to improve the livelihood.

Combining all these, the chapter provides an idea, as to how the TRUs survive through interdependence with the forest on a daily basis by following the rules and regulations put in place by them based on their traditional knowledge. The TRUs consist of different groups performing specific tasks like mouals (honey/wax collectors), bawalis (woodcutters) and golpata (*Nypa fructicans*) harvesters for earning their livelihood and each of these groups plays distinctive roles that have been explained in this chapter. The chapter has also briefly covered the people involved in crab culture. Finally, the role of TRUs in the conservation of the forest by maintaining its biodiversity through sustainable usage of resources has also been explored.

Traditional Rules and Practices Followed by TRUs While Harvesting Resources

The TRUs consciously keep the forest in high esteem as they regard it as a provider of their livelihood (see Titumir et al., 2020; Titumir & Paran,

2022a, 2022b, 2022c). Hence, they believe that it ought to be protected from all kinds of abuse and this is true for all generations. Therefore, based on their traditional knowledge they have put in place certain rules, which when followed in spirit ensures that resources are harvested with absolute care and love for nature in a harmonious way.

Rules Followed by Mouals (Honey/Wax Collectors)

Honey is an important non-timber product collected from the forest. It is in the month of April, May and June that the mouals (honey/wax collectors) enter the forest to collect honey from the honeycombs built by wild bees on the branches of trees. While collecting honey the mouals ensure that only a certain portion which is about two-thirds of the beehive is cut for honey collection. Rest of the beehive is left behind on the branches for the bees to return and resume their work. To ensure that young bees are not killed during the collecting process, the collectors squeeze the hive with their hand and refrain from using metal tools and to chase off the bees, smoke is produced using dry leaves and it is ensured carefully that the beehive is not put on fire. After one round of collection, the mouals revisit the same colony only after a month depending on the size of the colony and the flowering condition of nearby vegetation to stall overexploitation.

Rules Followed by Bawalis (Wood Collectors)

The bawalis (wood collectors) follow numerous guidelines to ensure sustainable harvesting of wood from the forest. They leave behind at least one stem in each clump of trees while collecting wood. Once the bawalis have harvested wood from a compartment, they will not use this compartment for harvesting in the following year. They cut wood on a seasonal basis to ensure adequate regrowth of plants. They usually cut wood from areas where it is plenty and they refrain from cutting young and straight trees.

Traditional Practices of Golpata (Nypa Fruticans) Harvesters

The golpata harvesters strictly refrain from collecting from an area more than once a year. Collection of golpata is stalled from June to September and this rule is sacrosanct as this is regarded as the fruition period of *Nypa*

Fig. 4.1 Traditional rules and practices followed by TRUs at a glance (*Source* Adapted from Titumir and Afrin [2017])

fruticans. As per their rules, only the leaves are cut which are roughly nine feet long and it is cut in a manner that the central leaf and the leaf next to it in each clump are left behind for regrowth. While collecting golpata, a rule that is strictly followed is to make sure that the flowers and fruits of the plant remain intact. They also ensure that young plants with only one utilisable leaf are not cut.

Figure 4.1 helps to classify the different groups of TRUs and portrays their tasks respectively. Each task is unique and can easily be distinguished from the other. Also, each task is carried out with utmost care and precision.

Cultural Beliefs of TRUs and Indigenous Communities

There are certain cultural beliefs in place which guide the activities and practices of the TRUs to a great extent since antiquity. This belief system significantly helps in conserving the forest resources and helps maintain the biodiversity. Over a period of time the locals and indigenous people in this region have been influenced by different religions and hence they consist of both Hindus and Muslims but their lives within the forest are guided by their ancestral belief system (Sarkar, 2010).

Bawalis

The bawalis consider that this tidal forest is a holy place that the Creator washes twice a day (high tide and low tide) to maintain its purity. Regardless of religion, they believe in the existence of 'Banabibi', who is the main Goddess of the forest including other Gods and Goddesses. They believe that their minds are calmed when they remain inside the forest. These cultural beliefs guide the primary task of woodcutting, and hence, they try to use the resources of the forest sustainably.

Mouals
The mouals also believe in Banabibi's existence and pray to her before entering the forest. While praying the mouals ask for good produce of honey and their safety, as many of them become victims of tiger attack each year as they roam the forest looking for honey. The Hindus also believe in the existence of another God for honey who is named 'Dakshina Roy', to whom they pray. The mouals consider the forest to be a holy place because of the existence of different Gods and Goddesses. Thus, they try to keep the forest safe and pure by following certain norms of not entering the forest at certain hours that might agitate the wildlife, refraining from defecating or smoking in the forest and entering the forest with clean hands by not carrying arms.[1] They also ensure to not hurt the forest, by using its resources efficiently. They consider honey to be a sacred food and are careful to not adulterate it.

Munda Community
The 'Mundas' are indigenous people, who were the first to settle near the forest area. They consider themselves to be a part of the forest. They consider that outsiders have destroyed the whole ecological and social harmony of the forest area as a result of which the Mundas want to keep their own land, forest and natural resources limited to them (Figs. 4.2 and 4.3).

Innovations in Livelihood Options

In addition to the traditional rules, practices and cultural beliefs mentioned earlier, the TRUs have diversified their livelihood choices by implementing their traditional knowledge and experiences (see Baten & Kumar, 2010; Kabir & Hossian, 2008; Titumir, 2011, 2015; Unnayan Onneshan, 2010). They regularly attempt to adjust to the changing conditions by deploying innovative management practices.

[1] Available at http://www.environmentandsociety.org/arcadia/bonbibi-religion-forest-sundarbans (last accessed March 2021).

Fig. 4.2 Community resource centre of the Munda indigenous people (*Photo* Unnayan Onneshan)

Fig. 4.3 A Drawing showing Glorious history of Munda community (*Photo* Unnayan Onneshan)

Innovative Techniques in Agriculture

The small and marginal local farmers developed innovative approaches in agriculture that can be adapted to local biophysical conditions while ensuring environmental sustainability. Due to climate change and the increasing salinity in the soil and water of the coastal region, farmers now sow rice in higher areas away from the coast. This practice has been adopted recently to lower the risk of saline water contamination and ensure optimum survival of the sown seed. Then, the seedlings are transplanted to the main agricultural land. Now the farmers cultivate rice at 8–12 inches high from the ground level, to save the crop from rising salinity in both soil and water. Technically, this saline contaminated rice straw is dissolved within a very short time if used as a roofing material. Hence, they allow those to be decomposed in the field which in turn adds organic matter, mainly nitrogen, to the soil to reduce saline intensity. This is favourable for the growth of their next crop. Moreover, the landless people grow vegetables on sheds or roofs, in frontyard or backyard of their houses.

Joint Cultivation of Crab and Duck

The TRUs often face restrictions entering the forest for resource collection. Hence, they have developed alternative sources of livelihood. A recently adapted practice is rearing of ducks jointly and that of crab culture in one farmland. This practice is by far found to be much profitable for the cultivators. However, intensive research is required to realise the full potential of these practices, in terms of both economic and environmental sustainability.

Community-Based Mangrove Agro Aqua Silvi (CMAAS) Culture

The CMAAS culture is referred to the practice of integrated cultivation of few mangrove faunal species such as crabs, oyster, variety of fish like shrimps, bhetki (Latescal carifer), etc., and floral species like golpata (*Nypa_fruticans)*, keora (*Sonerati_aapetala*), goran (*Ceriops_decandra*), etc., at the same time on available swampy land in brackish water. In addition to it, integrated cultivation of certain mangrove floral species like golpata and a few faunal species like tengra (*Mystus_tengara*), baila (Awaous_guamensis), tilapia (*Tilapia nilotica*), etc., are practised in freshwater swampy land. CMAAS culture is innovated as a substitute to the

Commercial Shrimp (CS) culture and has little or no adverse impact on the ecosystem (please see chapter 9 for details on CMAAS culture).

Marginalisation of TRUs and Degradation of Resources

As discussed in the previous chapter, the Sundarbans has been considered to be a Reserve Forest area since the British rule, along with other specific regions that are also regarded as Protected Areas. Hence, the TRUs have to ask permission to enter the forest for harvesting resources. These restrictions are presented as a step that would assist in minimising the degradation of forest resources and its depleting biodiversity. This model of protected area management has failed to achieve the expected outcome—both with respect to biodiversity conservation or in providing security to the livelihood of the TRUs.

Figure 4.4 shows the overall picture on the basis of the three pillars of access and benefit sharing, conservation and sustainable usage of the resources. It can be seen that access to benefit sharing causes restrictions on the rights of the TRUs and not for the influential political parties. This in turn results in the destruction of the biodiversity of resources, as shown in the diagram. Similarly, it is also seen from the diagram that the TRUs contribute by their knowledge and livelihoods in the conservation and sustainable use of resources to protect the biodiversity. But they cannot utilise their traditional knowledge in maintaining their livelihoods as they are marginalised in the management framework.

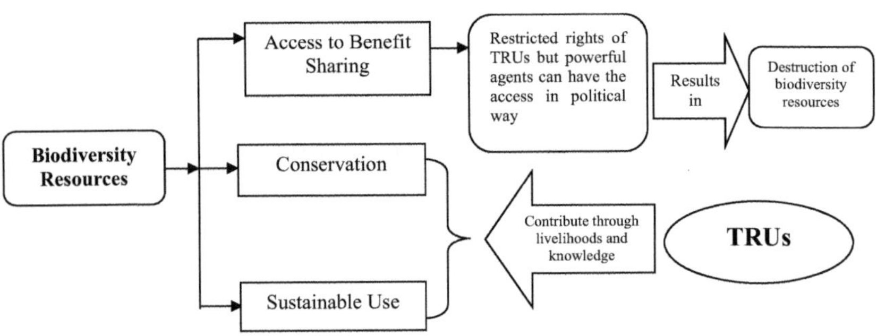

Fig. 4.4 Role of TRUs in the conservation and sustainable usage of biodiversity resources (*Source* Prepared by authors)

Conclusions

Throughout this chapter, the traditional knowledge and customary rules of the local forest people have been explored to establish these as significant means of conserving the forest's biodiversity. Among the TRUs dependent on the terrestrial resources of the forest, customs of three groups, that is the bawalis (golpata collectors), the mouals (honey/wax collectors) and the golpata harvesters have been presented in detail to showcase their knowledge and practices that are useful for the sustainable use of the forest resources. Here, great emphasis has been given to the ways, each of these groups deploys in carrying out their tasks in detail—starting from when and where they work to how they work. Their work reflects sustainable and proper use of resources and also helps reveal what is the actual and sensible way of carrying out these tasks. This is mainly true because these people live near the forest and their lives are linked to it. Hence, it can be said that they know better about the forest compared to anybody else.

The cultural beliefs of the TRUs act as foundations of their lives and practices. Summing up all these can help to depict how the TRUs have played vital roles in conserving the forest resources. They consider the forest to be protected from any sort of misuse and abuse for all generations. Doing so will preserve the resources which can be used for a longer time and in a proper manner. Also, the roles of the TRUs in the conservation and sustainable usage of biodiversity resources were shown with help of a diagram. This diagram revealed how one factor led to another and how all the factors are interlinked with each other. Furthermore, this chapter also provides some innovative techniques in agriculture that are used by the farmers. This is mainly to tackle the case of climate change and increase in salinity in the soil. In addition, the cultivation of duck and crab culture was also highlighted which revealed its profitability extents. Finally, the chapter depicts the CMAAS culture which took into account the flora and fauna that revealed how CMAAS culture was a substitute to the Commercial Shrimp (CS) culture.

The biodiversity of resources in this mangrove forest has been deteriorating by the day due to various anthropogenic pressures. These pressures have intensified with the advent of neo-liberalism that promotes market forces as the only strategy for accumulating wealth at the cost of nature. The commercial enterprises, both formal and informal, are highly organised in their extractions of resources and quite often are politically patronised and administratively supported. Therefore, restricting the TRUs from entering the forest cannot be the way forward towards stalling biodiversity degradation. Overall, the TRUs can vastly lead to the conservation and sustainable use of the biodiversity resources by means of their traditional knowledge and cultural beliefs.

References

Baten, M. A, & Kumar U. (2010, May 3–5). *Responses to the changes in the Sundarbans.* Paper presented at international conference on biological and cultural diversity, UNCBD, Montreal, Canada.

Kabir, D. M. H., & Hossain, J. (2008). *Resuscitating the Sundarbans: Customary use of biodiversity and traditional cultural practices in Bangladesh.* Unnyan Onneshan, BELA, Forest Peoples Programme and Nijera Kori, Dhaka.

Sarkar, S. C. (2010). *The Sundarbans: Folk deities, monsters and mortals.* Orient Blackswan.

Titumir, R. A. M. (2011). *Property rights, CSU and development.* Paper presented at the Expert Committee meeting on CBD article 8(j), CBD Secretariat, Montreal.

Titumir, R. A. M. (2015, November 6). IPLCs contribution to Aichi Biodiversity Target 10: a case illustrated through TRUs of Sundarbans in Bangladesh. Paper presented at side event on indigenous peoples and local communities' contributions to the implementation of the strategic plan for biodiversity 2011–2020, Montreal.

Titumir, R. A. M., & Afrin, T. (2017). Complementarities of human-nature well-beings: A case illustrated through traditional forest resource users of Sundarbans in Bangladesh. *Satoyama Initiative Thematic Review (SITR)—Volume 3*, United Nations University Institute for the Advanced Study of Sustainability (UNU-IAS).

Titumir, R. A. M., Afrin, T., & Islam, M. S. (2020). Traditional knowledge, institutions and human sociality in sustainable use and conservation of biodiversity of the Sundarbans of Bangladesh. In O. Saito, S. M. Subramanian, H. Hashimoto, & K. Takeuchi (Eds.), *Managing socio-ecological Production Landscapes and Seascapes for Sustainable Communities in Asia: Mapping and Navigating Stakholders, Policy and Action* (pp. 67–92). Science for Sustainable Societies. https://doi.org/10.1007/978-981-15-1133-2_5

Titumir, R. A. M., & Paran, M. S. (2022a). Human-nature cooperation for well-being: Community understanding on one health approach in the COVID-19 era in the Sundarbans. In M. Nishi, S. M. Subramanian, H. Gupta

(Eds.), Biodiversity-health-sustainability nexus in socio-ecological production landscapes and seascapes (SEPLS). Springer Nature.

Titumir, R. A. M., & Paran, M. S. (2022b). Ecosystem services and well-being in the Sundarbans of Bangladesh: A multiple evidence base trajectory. In R. Dasgupta, S. Hashimoto, & O. Saito (Eds.), *Assessing, mapping and modelling of mangrove ecosystem services*. Springer Sustainability Science Book Series (Forthcoming).

Titumir, R. A. M., & Paran, M. S. (2022c). Human-nature relationship for post 2020 biodiversity framework: An exploration into traditional knowledge, customary sustainable use and community-based innovation in the Sundarbans. In R. A. M. Titumir, Department of Development Studies (Eds.), *Fifty years of nation building: Political economy of Bangladesh's development*. Centennial Book Series Project of University of Dhaka. University of Dhaka (Forthcoming).

Unnayan Onneshan (UO). (2010). *Community based mangrove agro aqua silvi (CMAAS) culture: Promoting as a community adaptation tool and an alternative to commercial shrimp culture* (Research report). Unnayan Onneshan, Dhaka.

CHAPTER 5

Traditional Knowledge and Customary Sustainable Practices in Coastal and Marine Ecosystem

Rashed Al Mahmud Titumir and Mostafa Walid Pasha

Introduction

This chapter attempts to explore the existing customs and traditional fishing practices of the indigenous and local fishermen living along the coastal belt and the impact of such practices on the marine ecosystem. Through proper study, during the course of the chapter, it is established that traditional knowledge and the old customs promote sustainable fishing practices and are environment friendly as they help in the conservation of marine resources. The sustainability and conservation of natural resources depend on the inherent human practices deployed in using such

R. A. M. Titumir (✉)
Department of Development Studies, University of Dhaka, Dhaka, Bangladesh
e-mail: rt@du.ac.bd; rtitumir@unnayan.org

R. A. M. Titumir · M. W. Pasha
Unnayan Onneshan, Dhaka, Bangladesh

R. A. M. Titumir (ed.), *Sundarbans and its Ecosystem Services*, Sustainable Development Goals Series,
https://doi.org/10.1007/978-981-19-3000-3_5

resources. Conscious social interaction in tandem with nature promotes natural harmony and thereby ensures conservation.

The Traditional Resource Users (TRUs) follow and maintain certain rules which have been passed on to them through generations (see Titumir et al., 2020). These rules still exist and are prevalent because of constant practice. Years of knowledge, practice and numerous conventions have gone into the formulation of these rules. The reason that these rules have been formulated is to ensure the collection of resources efficiently by ensuring that none of the resources are harmed or get depleted in the process of collection. As mentioned earlier, the TRUs consider the forest as an extended part of their family or as a close relative and are also dependent on it for their livelihood and hence exploiting it would only harm them in return (Titumir & Paran, 2022a, 2022b, 2022c). These age-old customs dictate their mode of fishing and ensure that marine resources are properly conserved along the coastal belt. These customs restrict use of certain fish catching gear and traditional knowledge helps the TRUs to understand the habitat by observing the colour of the water, feeling wind direction, mud, etc., that have evolved with the passage of time.

Nevertheless, the opinion of the TRUs is not taken into consideration while discussing the conservation process in the Sundarbans. Their participation is not encouraged while promoting new development activities. If by chance their participation is sorted, it is ensured that their opinion based on years of traditional knowledge is not included. However, people who have little or no knowledge about the forest and are from outside dictate the conservation norms as most of them have developed a network of clientelism involving the police, forest department employees and local politicians (Titumir et al., 2019). This profit-seeking group has crippled the local and indigenous people hampering the existing traditional sustainable practices that encouraged conservation. The human sociality and close kinship that the TRUs share with the forest and their cooperative sustenance gets interrupted by these external negative agencies as a result of which the TRUs suffer both economically and socially.

This chapter includes the study of both traditional and non-traditional methods after gathering information from field studies. Proper references from corroborative literatures have also been included to strengthen the argument to establish the fact that traditional knowledge is necessary for the conservation of the forest resources. During the course of this study, it was found that shrimp farming is working as a catalyst and providing impetus to the destruction of the marine ecosystem. Local and indigenous fishermen employing traditional fishing practices showed concern over the increasing number of *gher* (enclosures) used for farming shrimps. The rising number of such *ghers* has led to pollution and is also destroying many other species of fish as harmful and poisonous chemicals are used for improving shrimp production. Most of these *ghers* are owned by affluent people who are oblivious about conservation and are profit-oriented.

Rules of Fishing in the Water Bodies: Traditional vs Non-traditional Perspective

The fishing practices of the TRUs are different from the modern ones. These differences are the ones that tread the thin line separating sustainable methods from the unsustainable ones that cause harm to the prosperity of aquatic resources in the forest.

TRUs' Perspective

The TRUs have been relying on traditional knowledge bank while collecting resources from the forest. They consider the forest to be their mind and hence they believe that harming the forest is like harming themselves. Based on this traditional knowledge bank, specific rules have been formulated to ensure the collection of forest resources in a sustainable manner. A local woman, Ambia Khatun, said,

> We do not do anything that harms the forest. If we harm it, we will not be able to survive. When we go for fishing, we consider the greater fish resource of the forest. That's why we do not use any dangerous net or chemicals while going for fishing.

Traditionally fishermen follow specific techniques during the fishing period. On Fridays, they barely go for fishing. They use locally designed tools for fishing such as locally made nets which are not harmful to carp variety of fishes. The rules that are in place traditionally prohibit the use of harmful tools like current nets, *bainjaal*, etc. Earlier, the fishermen of Koyra region had free access to the forest but during the field research, it became evident that now they don't have free access. They have to take a pass from the forest department to enter the forest for fishing. In addition to it, the traditional fishermen have to face the aggression of the powerful groups involved in using hazardous fishing-nets drawn by engine-operated trawlers. In case the people operating the trawlers get caught by personnel of the forest department, they save themselves by bribing the officials. Thus, they form a well-negotiated network of clientelism involving the police, local politicians and personnel of the forest department. Abu Musa, a local fisherman said,

> Forest department only knows how to harass us. On the other hand, they do not take any steps against the opportunist groups who are continuously snatching the resources without any consideration of the conservation aspect. These people use bainjaal, chorjaal and other dangerous nets. As a result, small fish and fish pollen get destroyed. We are losing our fish resources. We have been suffering both financially and environmentally.

Customary Rules Followed by Jele (Traditional Fishermen)

Traditional fishermen, also known as the *Jele* in the local language are well aware of the fact that catching fry (small fish) will eventually lead to depletion in the number of fish in the water, so they avoid catching small fish. The traditional fishermen refrain from using polyethylene netting *jal* with a very small mesh size to minimise waste and discard during fishing which is detrimental to the marine ecosystem. Rather they use selective, environment-friendly nets locally known as *behundijaal* (bag nets) or *charpaata* and *khaalpaatajaal* (stake nets), developed and customised scientifically over ages to benefit the unique waterscape. Traditionally fishermen use big-meshed nets for rivers and small-meshed nets for closed water bodies. All species of fish are not captured and fishing during the breeding or spawning season is avoided to arrest resource depletion. Just like other groups of TRUs, *jele* or the people belonging to the fishing community also consider the forest to be a sacred place. The specific reasons behind such considerations are (a) No sinful act takes place within the forest premises as it is devoid of human, (b) Forest is the abode of different Gods and Goddesses and (c) They firmly believe that the forest is a gift from the Gods to them.

In Table 5.1 both traditional and non-traditional rules are mentioned to provide a better understanding of comparative outcome from both kinds of uses of coastal and marine resources.

From Table 5.2, it is certain that measures taken by traditional fishermen are conducive to the sustainability of the aquatic resource ecosystems of the forest. Contrarily, a particular rent-seeking group has

Table 5.1 Traditional vs. non-traditional rules of the fish resource usage

Item	*Traditional rules*	*Non-traditional rules*	*Remarks*
1. Spatial restriction	Temporary closure of 1–2 weeks to get a considerable amount of fish in certain areas which get abundant with more species of fish after the closure period ends	No spatial restrictions are maintained; extraction of fish is happening constantly from all water bodies	Fishing opportunities get decreased and some important species of fish are getting reduced
2. Temporal restriction	Normally they do not gofishing on Friday. Alongside, some small period restrictions are imposed when getting more fish becomes necessary	No temporal restrictions are maintained	Because of incessant fishing by the outsiders, the fish resource is getting reduced gradually
3. Gear restriction	They avoid using fishing techniques that destroy the carp of fish	Detrimental nets like '*bainjaal*' and '*chawrjaal*' are used for fishing	Because of the pernicious techniques of fishing, larvae, carp and fish egg get diminished
4. Effort restriction	There is a jointly negotiated fishing opportunity using traditional nets like '*chawrpata*' and '*Khalpata*'	No particular restriction of using techniques, spaces and who can use or not	Outsiders are extracting as much fish as they can conniving at the rules and regulations
5. Species restriction	Restrictions are put during the ovulation period of certain species like *paissha, dadne, vetki (Latescalcarifer)* and other fish	During the use of nets like '*chawrjaal*', '*bainjaal*' and some foreign nets, fish including their carp get caught in the net	Carps die which results in fewer fish on the canals, rivers and coastal areas
6. Catch restriction	Only consumable species of fish are caught	Most of the species of fish are caught	Natural biodiversity gets disrupted and harmed
7. Water colour	When the water gets reddish, they predict more *hilsha (Tenualosa ilisha)* production in the coastal region. Amount of fish increases in turbid water. Number of fish gets decreased in polluted water	Outsiders are devoid of this idea mostly	Due to use of engine boats and trawlers by the outsiders, water gets polluted which kill many species of fish and hamper reproduction

(continued)

Table 5.1 (continued)

Item	*Traditional rules*	*Non-traditional rules*	*Remarks*
8. Weather, wind direction and current	Fish increase during high tide. Wind from both southern and south-east direction imply increase in amount of fish	Non-traditional resource users do not follow this perspective. Many of them have learnt the traditional rule recently	More extraction of fish is being occurred as outsiders utilise the derived knowledge of the fishermen
9. Lunar periodicity	During the full moon amount of fish increases. Hence it is wise to wait for the full moon to catch more fish	They go fishing all the time regardless of full moon and new moon	More extraction occurs and carps get killed
10. Sediment and topography	The high amount of siltation is dangerous to the topography of the water bodies. They also detect the type of mud through their finger and can predict most catchable fishing sites	Non-traditional forest users cut trees incessantly which creates more sedimentation	Due to more sedimentation marine and coastal ecosystems gets disrupted and fishing sites get decreased
11. Traditional celestial navigation	They use the traditional method of static (e.g. location of the site) information to calibrate the relative position of the fishing boat	Non-traditional resource users mostly wear wristwatch to get to the direction of the fishing site	The traditional method is more reliable because of the years of experienced practice
12. Birds	They consider birds like pigeons, egrets, common tern, kingfisher as sacred. These birds also help them find a suitable place for fishing	No particular affiliation with particular species of birds	Birds are also killed by hunters which damages the biodiversity of the forest
13. Fishing sites	They divide fishing sites based on different canals like '*Bustamkhal*', '*Keora kata*', '*Kodal kata*', etc.	They are likely to go fishing where they can extract more fish	Due to the encroachment in almost every water body, traditional fishermen are getting deprived of fish than before

Source Compiled by authors

Table 5.2 Traditional rules followed by different communities

Item	*Traditional rules*
1. Spatial restrictions	For numerous months areas are closed for fishing temporarily
2. Temporal restrictions	Fishing is restricted on some specific days, weeks or months to increase the supply of fish in the following period
3. Gear restrictions	Prohibition of some certain tools or technologies that are detrimental to the fish resource
4. Effort restrictions	Fishing is restricted to a particular family or is community-based on catching, gear and species restriction
5. Species restrictions	Catching of some specific species of fish is forbidden. It sometimes occurs due to an increase in the supply of valuable fish and also occurs based on some rituals where some fish are regarded as bliss for the local people in some certain perspectives
6. Catch restrictions	In some regions harvesting of fish is limited to minimise wastage. Quite often social norms work as a catalyst in this context
7. Water colour	Fishermen find six types of water mass from the Bay of Bengal: turbid, clear, light green, green, bluish and brackish. For example, if the colour of the water is red, fishermen can smell Hilsa mucus in the river
8. Weather, wind direction and current	Wind from the north and east to west is considered good for fishers. The southerly wind that blows from mid-March to mid-July brings *bish (poison)* as per the fishermen of marine areas of Bangladesh
9. Lunar periodicity	According to the fishermen, tidal height during a full moon is stronger than that of a new moon. Both Muslim and Hindu fishermen of Bangladesh believe that in every Bengali lunar month there are two phases of high tide. They assume that in each fortnight there are 10 days of *jo* (high water periodicity) and five days of *dala* (low water oscillation)

(continued)

Table 5.2 (continued)

Item	*Traditional rules*
10. Sediment and topography	Fishermen of the marine areas of Bangladesh found that fish always swim along the *dhair* (edge) of the seamounts. Big fish that can combat higher velocity underwater currents are found along the deeper portion of the *khari* (bed of the valley). Fishermen also concentrate on the pattern of mud and categorise them in classes like sandy, gravel-rich, muddy and slightly silty mud
11. Traditional celestial navigation	Artisanal fishermen do not refer to compass for navigation. Very few of them have a wristwatch. However, they use practical skills like identifying location according to the position of the stars. They usually keep track of mental calculations and use their experience in this field
12. Birds and animals	In some areas, pigeons, black-headed gulfs and some other hunter birds are regarded as indicators of getting big fish as these birds do not catch small fish. Fishermen of the south-west of the Bay of Bengal use otters as a fish catcher. Otters are very clever as they can catch quite a good number of fish in a short duration
13. Fishing sites	Fishermen detect some specific fishing sites for catching fish. In this case, traditional and social norms also determine their fishing approach, e.g., caste-based Hindu fishermen have an age-old territorial management system called *faar* by which fishing entitlements over certain fishing spots are granted to community members

Source Compiled by authors based on Cinner and Aswani (2007) and Deb (2015) come up who do not conform to the socially established rules and restrictions practised by the TRUs. As a result, coastal and marine resources get destroyed and the natural biodiversity of faunal species of the water bodies get interrupted.

There are many types of restrictions that can ensure the sustainable use of the resources. These include spatial restrictions, temporal restrictions, gear restrictions, effort restrictions, catch restrictions, species restrictions, compliance and enforcement and so on. Apart from Bangladesh, other countries also use traditional knowledge for the conservation of natural resources. There are examples of temporary reef closures before religious ceremonies, to replenish the supply of fish and other invertebrate species (Aswani & Weiant, 2004; Hickey, 2006; Hviding, 1989). They also impose reef closure after the death of a landowner or village chief and after a marriage or birth (Hickey, 2006; Macintyre & Foale, 2007; Ravuvu, 1983; Wright, 1985). In Maluku, Indonesia, specific community leaders use customary management to ban cyanide fishing associated with the live reef food fish trade (Thornburn, 2001). In certain areas, particular groups have the right to use a specific fishing technique and the same is restricted for others (Carrier & Carrier, 1983). Currently, coastal groups in the Roviana Lagoon, Solomon Islands, are not only excluding non-owners of fishing areas/waterbodies from using restricted technologies such as spearfishing at night but are also constantly demanding that inclusive stakeholders avoid using these fishing methods (Aswani & Hamilton, 2004). There are also restrictions on who can catch particular species of fish and which method they should follow and in which area they can go for fishing (Cinner & Aswani, 2007). Apart from such restrictions, other rules are also put in place to ensure the sustainable use of the forest resources based on the colour of the water, wave, wind direction and current, lunar periodicity, sediment and topography, traditional celestial navigation, presence of specific birds and animals, mangroves, fishing sites, etc. The traditional rules followed by the traditional resource users, compared to the non-traditional users are shown in Table 5.2.

Table 5.3 shows fishermen's decision-making process in the sea on different parameters and observations. It takes into account various major factors and takes necessary actions based on supporting observations and decision-making processes.

Table 5.3 A few examples of fishermen's decision making process in the sea based on different parameters and observations

If (major factor)	*And/or (supporting observations)*	*Then (decision making)*	*Actions*
Water colour is reddish	– Frothy and sticky substances are present – On the surface – Hilsa mucous is smelled	– Possibility of Hilsa shoals around	– Encircle the area with gill net
Two distinct water masses merge	– Water current is within limit of tolerance for gear – Small fish and shrimp seen jumping	– Possibility of catching both brackish water and marine species	– Set ESBN or MSBN in the area
Southerly violent wind is blowing	– Fatty Harpodonneherius is abundant in the net – Dark cloud – Wind seems to burn skin	– Cyclone is imminent – Consult with fellow fishers	– Wind up the gear and steer towards the coast – Set gear close to the coastline after cyclone
Sun is bright	– Wind is calm – See the movement of small fish – See some zooplankton in a glass of water; bait is ready – Hunter birds around	– Ideal for shoals of Polynemus indicus, Hilsa sps, sharks, and Eleutheronema tetradactylum	– Set the hooks 5–8 m below surface – Set gill net with weight at 4–7 m depth
New moon darkness	– Close to mangroves – Shrimp's breeding season – Bottom is smooth	– Ideal for targeting migrating species	– Set gill net reaching bottom or encircle the area by joining nets

Source Adapted from Deb (2015)

Practices of Fishing Followed by the Fishermen: Traditional vs Non-traditional Perspective

Other than these several rules, traditional resource users, especially fishermen keep up with several practices of fishing that are harmless to the aquatic biodiversity of the forest. This process helps flourish their human sociality.

TRUs' Perspective

TRUs believe that proper appropriation of aquatic resources like fish, crab, shell, shrimp and other faunal species will help preserve the resources of the coastal ecosystems. However, continual expropriation of resources leads to the loss of resources which is mainly done by the powerful groups. Fatema Begum, one of the female fishers, expresses her grief,

> Because of a notorious group of people and also another group named 'black party', we are losing fish day by day. Both forest and the forest people are bearing the loss. They negotiate with each other, bribe the police and the personnel of the forest people. They should be brought under punishment.

The nets used by the local fishermen do have big holes, to ensure the carps of fish do not get stuck. They make their boats with *Goran* and *Sundari* mostly. On the other hand, the outsiders or the group of people who try to expropriate as much as possible fish use dangerous nets which have very little holes that captivate the carps of fish. Under these conditions, fish get killed and reproduction gets hindered which in the long run reduces the number of fish. Table 5.4 shows the comparative provision of traditional and non-traditional resources.

Fishermen of other regions including the Indian part use traditional nets which are helpful for the natural breeding of the fish. They usually use traditional handmade nets. Most importantly, these nets do not capture small fish and larvae and the reproduction of fish remains undisturbed and balanced. Table 5.5 provides a proper understanding of the nets used by traditional fishermen and the type of wood they use to make their boats.

The Negative Impact of Shrimp Cultivation in Coastal and Marine Ecosystems: TRUs' Perspective

Local fishermen are concerned with shrimp cultivation which is going up by the day. The fishermen complained that they face severe problems

Table 5.4 Traditional vs. non-traditional practices of fishing

Item	*Traditional practice*	*Non-traditional practice*	*Remarks*
1. Netting the fish	They use '*rocketjaal*' and '*berjaal*' mostly to catch fish	They use '*bainjaal*' and '*chawrpata*' and some foreign nets to catch fish	Due to the nets which are used by the outsiders, carps and larvae get caught and eventually killed. This lowers the rate of fish reproduction
2. The nets	The nets have medium to big holes which are mostly used to catch medium to big fish. The length of '*berjaal*' varies from 4–5 m to 90–100 m and the width is around 7–8 m	The nets they use are usually more than a hundred metres long and contain very tiny holes	As the nets of the non-traditional users are dense which contain very tiny holes, carps and larvae get caught and killed
3. Boats	They build a boat using *sundari* (Heritiera fomes) mostly. The boats are traditionally called '*dinginouka*'	Most of the non-traditional resource users use engine boats and trawlers from different kinds of wood like *gewa* (*Excoecariaagallocha*), *goran* (*Ceriopsdecandra*) and *sundari* (*Heritierafomes*)	As more outsiders are coming in for fishing, they are using more engine boats that cause pollution in the water, which in turn destroys the ecosystem of the fish

Source Compiled by authors

due to ever-increasing load of shrimp girdles. In addition to it, fish as a resource is getting affected and killed because of the high use of chemicals involved in shrimp cultivation. The imported shrimp fries (larvae) contain diseases that are passed on to other fishes.

However, instead of putting in efforts to stall such hazardous practices the authorities have colluded with the shrimp cultivators. The locals revealed that local politicians are hand in glove with local administration in establishing innumerable shrimp enclosures to earn huge profits within a short period of time. These people poison the shrimp, in addition to

Table 5.5 Traditional rules followed by fishermen

Item	*Traditional rules*
Netting the fish	Fishermen of the Indian Sundarbans use gillnet, *gansha, beonti* and *berjaal*, etc. for fishing
The net	The nets used by the fishermen of the Indian Sundarbans are a hundred to thousand metres long. The mesh size is usually between 10 to 25 cm
Boats	In the Indian part of the Sundarbans, boats are mostly made of *passur* (*X. mekongensis*), *sal* (*Shorearobusta*), *keora* (*Sonneratiaapetala*), Bharani etc. The boat ribs are mainly made of babla (*Vachellianilotica*)

Source Compiled by authors based on Chacraverti (2014)

poisoning the other water bodies, to get a big number of fish at a time. Consequently, two horrific problems arise: one is the killing of fish which also includes carps, while the other is the disastrous pollution of water. One of the local fishermen, Shajahan Ali, conveyed,

> If the forest department were strict enough to the influential group and at least executed the existing laws, this phenomenon would not rise. The powerful group is being indulged because all of them are shareholders of profit, and we are losing capital return day by day.

Figure 5.1 shows the negative effects of shrimp cultivation on natural fish resources. It shows how harmful actions such as use of poisonous chemicals, the imposition of power by rent-seeking groups and lax enforcement of laws can hamper shrimp cultivation by the loss of fish resources, water pollution and damage of the mangrove ecosystem.

Local fishermen concluded that through appropriate rules and breaking the cycle of the vicious network of the rent-seeking group, unethical cultivation of shrimp will be mitigated along with the reduction of using poisonous chemicals.

Shrimp aquaculture in Bangladesh depends greatly on the input of artificially formulated feed and the application of agrochemicals, antibiotics and disinfectants.

The most common products used in shrimp aquaculture are fertilisers for the enhancement of natural feed and liming material for water and soil control. Due to antibiotic use in shrimp cultivation, bacteria get antibiotic-resistant which causes more harm to shrimp and also fish via

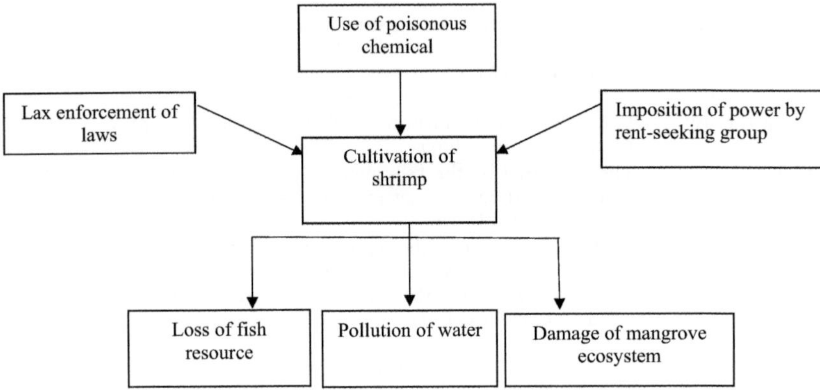

Fig. 5.1 Negative effects of shrimp cultivation on the natural fish resources (*Source* Prepared by authors)

severe diseases (Paul & Vogl, 2011). Several kinds of impacts are stated in Table 5.6.

Conclusions

It is evident that in conserving the natural environment and biodiversity, traditional practices contribute greatly. Integrating human behaviour into the formal model of natural resource management is still a big challenge (Janssen and Jager, 2000; Fulton et al., 2011; Milner-Gulland, 2012; Schlüter et al., 2012). Customary practices must be incorporated while starting informal projects by ensuring better participation of the forest people for the purpose of conservation and sustainable use of the forest.

The traditional fishermen have derived a perpetual relationship with the mangrove forest. They believe that rules and practices should help to conserve the natural aquatic biodiversity of the coastal and marine ecosystems. They value the fish resources not only from the perspective of economic opportunity and consumption fulfilment but also consider them as their life. Through the maintenance of traditional rules and norms, a long-lasting relationship has been formed where human sociality has been prioritised instinctively. Traditional fishermen convey that, due to irregular marketisation of fish and constant establishment of shrimp enclosures, fish production is negatively affected. Their natural breeding process is

Table 5.6 Impact of shrimp farming in marine and coastal regions

Sphere of impact	*Impact*	*Remarks*
Environmental	Destruction of the mangrove ecosystem	Mangrove destruction occurs due to shrimp aquaculture. Wetlands are converting to ponds/gher for shrimp farming
	Land degradation	Crop production declines due to salinity intrusion and the soil becomes fragile
	Loss of capture fishery stock	Farm owners force fishermen to catch wild spawn from estuary or coasts that causes losses as other fish and carps are caught indiscriminately
	Seed supply	Collection of wild shrimp seeds from the natural source of aquaculture operations directly affect other fish by lessening seeds for them
	Sedimentation	The sediment which is often released to waterways leading to the sea, from the girdle of shrimp is used to make new dykes or to maintain existing dykes. Maintenance of dykes damages the quality of water through wastage creation
	Pollution	In the case of intensive shrimp farming water changes every four to six days and through sluice gates the water gets exchanged with the main water flow which creates wastage of the girdle of shrimp in the natural water body
	Saltwater intrusion	Due to the continuous shrimp farming situation, saltwater increases in land areas which at one passage of time weakens the soil and decreases the fertility of the cultivable land

(continued)

Table 5.6 (continued)

Sphere of impact	*Impact*	*Remarks*
	Danger of imported fry and genetic alteration	Very often fries are imported without quarantining which may lead to a fusion of disease carrier fries with other fries, along with fish during the water discharge
	Diseases	In the case of intensive shrimp farming, shrimps become vulnerable to diseases due to alterations of psycho-chemical factors like dissolved oxygen, temperature, pH, etc.
	Use of biological and chemical products in shrimp aquaculture	The use of chemicals in shrimp culture, make other species of fish vulnerable to diseases and pollution of water. It also hampers the breeding process of other fish
Social	Imposition of power provision in shrimp aquaculture	Influential people such as businessmen, government officers and local politicians enforce their power to establish girdle of shrimp for profit
Economic	Increase in the cost of production	The increase in inputs like pellet feeds, medicine, fuel, etc. makes shrimp farming crucial for general fishermen as it lowers the return to production cost
Institutional	Lack of enforcement of laws and policies	Laws and policies are not correctly implemented which causes a monopoly of power for a particular group of people

Source Prepared by authors based on Paul and Vogl (2011)

being impeded which tends to cause a shortfall in fish catch. Due to the massive profit-seeking mentality of particular groups, alienation of people from the natural aquatic ecosystems is taking place. As a result, commodity fetishism has risen. To accomplish the self-interest of the rent-seeking group, they are extracting fish as much as they can. They are not bothered about the conservation and future condition of fish resources. The circumstances become more complex when the negligent administration of laws and policies is observed. Forest department employees are charged for being partial and taking bribes. They do it on a negotiation basis with the profit-seeking group who are typically powerful people, having a political position in the locality. A triangular network within these institutions has developed, where everyone is a shareholder of the profit. Thus, institutional fragility has come forward.

Hence, the first and foremost thing to be done is to reconstruct the institutions. Second, strict laws and rules ought to be generated, along with its proper establishment and implementation. Third, traditional rules and practices of the traditional fish resource users should be taken into consideration, in the co-management process. The TRUs fret about being excluded in the formal conservation process. Even if they are included, their words are not taken into account. Therefore, it is necessary to ensure their participation.

Acknowledgements Some insights of the chapter draw on 'The Sundarbans is our mind: an exploration into multiple values of nature in conversation with traditional resource users (TRUs)', published by United Nations University Institute for the Advanced Study of Sustainability, Tokyo, Japan, 2019, in *Understanding the multiple values users associated with sustainable use in socioecological production landscapes and seascapes (SEPLS)*, Satoyama Initiative Thematic Review, vol. 5.

References

Aswani, S., & Hamilton, R. J. (2004). Integrating indigenous ecological knowledge and customary sea tenure with marine science and social science for conservation of Bumphead Parrotfish (Bolbometoponmuricatum) in the Roviana Lagoon, Solomon Islands. *Environmental Conservation, 31*, 69–83.

Aswani, S., & Weiant, P. (2004). Scientific evaluation in women's participatory management: Monitoring marine invertebraterefugia in the Solomon Islands. *Human Organization, 63*, 301–319.

Carrier, J., & Carrier, A. (1983). Profitless property: Marine ownership and access to wealth on Ponam Island, Manus Province. *Ethnology, 22*, 133–151.

Chacraverti, S. (2014). *The Sundarbans fishers: Coping in an overly stressed mangrove estuary*. International Collective in Support of Fish Workers, Chennai, 73–92.

Cinner, J. E., & Aswani, S. (2007). Integrating customary management into marine conservation. *Biological Conservation, 140*(3), 201–216.

Deb, A. K. (2015). Something sacred, something secret: Traditional ecological knowledge of the artisanal coastal fishers of Bangladesh. *Journal of Ethnobiology, 35*(3), 536–565.

Fulton, E. A., Smith, A. D. M., Smith, D. C., & van Putten, I. E. (2011). Human behavior: The key source of uncertainty in fisheries management. *Fish and Fisheries, 12*(1), 2–17.

Hickey, F. R. (2006). Traditional marine resource management in Vanuatu: Acknowledging, supporting and strengthening indigenous management systems. *SPC Traditional Marine Resource Management and Knowledge Information Bulletin, 20*, 11–23.

Hviding, E. (1989). Keeping the sea: Aspects of marine tenure in Marovo Lagoon, Solomon Islands. In K. Ruddle & R. E. Johannes (Eds.), *Traditional marine resource management in the Pacific Basin: An anthology* (pp. 7–44). UNESCO/ROSTSEA.

Janssen, M. A., & Jager, W. (2000). The human actor in ecological—Economic models. *Ecological Economics, 35*(3), 307–310.

Macintyre, M., & Foale, S. (2007). Land and marine tenure, ownership and new forms of entitlement on Lihir: Changing notions of property in the context of a goldmining project. *Human Organization, 66*, 49–59.

Milner-Gulland, E. J. (2012). Interactions between human behaviour and ecological systems. *Philosophical Transactions of the Royal Society of London*, Series B, *367*, 270–278.

Paul, B. G., & Vogl, C. R. (2011). Impacts of shrimp farming in Bangladesh: Challenges and alternatives. *Ocean and Coastal Management, 54*, 201–211.

Ravuvu, A. D. (1983). *VakiiTakui*. University of the South Pacific.

Schlüter, M., Mcallister, R. R. J., Arlinghaus, R., Bunnefeld, N., Eisenack, K., Hoelker, F., Milner-Gulland, E. J., & Mueller, B. (2012). New horizons for managing the environment: A review of coupled social-ecological systems modeling. *Natural Resource Modeling, 25*, 219–272.

Thornburn, C. (2001). The house that poison built: Customary marine property rights and the live reef food fish trade in the Kei Islands, Southeast Maluku. *Development and Change, 32*, 151–180.

Titumir, R. A. M., & Paran, M. S. (2022a). Human-nature cooperation for well-being: Community understanding on one health approach in the COVID-19 era in the Sundarbans. In M. Nishi, S. M. Subramanian, H. Gupta (Eds.), *Biodiversity-health-sustainability nexus in socio-ecological production landscapes and seascapes (SEPLS)*. Springer Nature, Singapore.

Titumir, R. A. M., & Paran, M. S. (2022b). Ecosystem services and well-being in the Sundarbans of Bangladesh: A multiple evidence base trajectory. In R. Dasgupta, S. Hashimoto, & O. Saito (Eds.), *Assessing, mapping and modelling of mangrove ecosystem services*. Springer Sustainability Science Book Series (Forthcoming).

Titumir, R. A. M., & Paran, M. S. (2022c). Human-nature relationship for post 2020 biodiversity framework: An exploration into traditional knowledge, customary sustainable use and community-based innovation in the Sundarbans. In R. A. M. Titumir (Ed.), *Fifty years of nation building: Political economy of Bangladesh's development*. Centennial Book Series Project of University of Dhaka. Department of Development Studies, University of Dhaka (Forthcoming).

Titumir, R. A. M., Afrin, T., & Islam, M. S. (2020). Traditional knowledge, institutions and human sociality in sustainable use and conservation of biodiversity of the Sundarbans of Bangladesh. In O. Saito, S. M. Subramanian, H. Hashimoto, & K. Takeuchi (Eds.), *Managing socio-ecological production landscapes and seascapes for sustainable communities in Asia: Mapping and navigating stakeholders, policy and action* (pp. 67–92). Science for Sustainable Societies. https://doi.org/10.1007/978-981-15-1133-2_5

Titumir, R. A. M., Paran, M. S., & Pasha, M. W. (2019). The Sundarbans is our mind: an exploration into multiple values of nature in conversation with traditional resource users. In UNU-IAS & IGES (Eds.), *Understanding the multiple values associated with sustainable use in socio-ecological production landscapes and seascapes (SEPLS)* (Vol. 5, pp. 97–117). Satoyama Initiative Thematic Review. United Nations University Institute for the Advanced Study of Sustainability.

Wright, A. (1985). Marine resource use in Papua New Guinea: Can traditional concepts and contemporary development be integrated? In K. Ruddle & R. E. Johannes (Eds.), *The traditional knowledge and management of coastal systems in Asia and the Pacific* (pp. 79–100). UNESCO.

CHAPTER 6

Multiple Values of Ecosystem Services and Human Well-Being

Rashed Al Mahmud Titumir, Md. Shah Paran, and Mostafa Walid Pasha

Introduction

This study attempts to categorise the Ecosystem Services (ES), their significance and examines the way ES contributes to human well-being. The chapter explores human well-being based on the contribution from ES and at the same time, it tries to understand in detail about better upkeep of nature. It demonstrates the appropriation of nature (conservation, restoration, sustainable uses, access and benefit-sharing) instead of expropriation (anthropogenic pressures) nurtures a symbiotic relationship between humans and nature that accentuates the quality of life.

R. A. M. Titumir (✉)
Department of Development Studies, University of Dhaka, Dhaka, Bangladesh
e-mail: rt@du.ac.bd; rtitumir@unnayan.org

R. A. M. Titumir · Md. S. Paran · M. W. Pasha
Unnayan Onneshan, Dhaka, Bangladesh

R. A. M. Titumir (ed.), *Sundarbans and its Ecosystem Services*, Sustainable Development Goals Series,
https://doi.org/10.1007/978-981-19-3000-3_6

The benefits from the ecosystems appear as multiple goods and services, the ecosystem services, that contribute to human well-being (Díaz et al., 2006; Layke et al., 2012; Millennium Ecosystem Assessment, 2005a, 2005b; van Oudenhoven et al., 2012). These benefits include food, medicines, raw materials, regulation of climate, provision of productive soils, disease prevention, clean water, as well as recreational, cultural and spiritual benefits (Banzhaf & Boyd, 2007; Costanza et al., 1997; Daily, 1997; Millennium Ecosystem Assessment, 2005a; Wallace, 2007). Ecosystems provide goods such as food crops, seafood, timber, biomass fuels, natural fibre, pharmaceuticals and geological resources. It also provides industrial products and services including waste management, cleansing, recycling and renewal, and intangible aesthetic and cultural benefits (Bengtsson, 1997; De Groot et al., 2002; King et al., 2000). Major categories of ES involve provisioning services, regulating services, cultural services and supporting services (Millennium Ecosystem Assessment, 2005a).

The supporting services are goods and other services which sustain various aspects of human well-being. The regulating functions affect human well-being in multiple ways, for instance, the purification of air, freshwater, reduced flooding or drought, stabilisation of local and regional climate, etc., that control the range and transmission of certain diseases, including vector-borne diseases. Cultural services such as totemic species, sacred groves, trees, scenic landscapes, geological formations or rivers and lakes influence the aesthetic, recreational, educational, cultural and spiritual aspects of human experience. On the other hand, supporting services are necessary for sustaining each of the three other services that indirectly affect human well-being (Millennium Ecosystem Assessment, 2005a).

Thus ES contributes to human well-being in numerous ways. Unfortunately, the services from the ecosystems are extremely undervalued, since most of these services do not have a price tag to trade them in the market. Hence, it is difficult to estimate the true value of ES (Daily et al., 1997). Moreover, market forces are oblivious of the role that ES plays as it is not priced and hence the many benefits of nature are neglected (Titumir et al., 2019). The value of nature can hardly be understood only through monetary valuation. Conceptualising value, consider a 'plurality of worldviews', of nature because people differ in how they understand the word 'value' and how they attribute importance to nature (González-Jiménez et al., 2018). Similarly, there are different scientific concepts of value—intrinsic, instrumental and relational values that cannot be conceptualised only by monetising the market mechanism. Values of nature are subjected to personal experiences, traditional knowledge bank, and interdependence of human–nature relationships, perceived within specific paradigms and methodologies, which is considered to be a crucial way of valuing nature (González-Jiménez et al., 2018).

Mainstream economics considers natural resources as commodities, for which it calculates the value of nature only in monetary terms. Valuing nature this way largely neglects the intrinsic value of the resources. Also, in this process, the resources are regarded exploitable for human use that leads to their depletion. Nature can be conceptualised in different ways (for example, the environment, Mother Earth, natural resources, natural capital). People refer to different types of values concerning nature (for example, intrinsic, instrumental, aesthetic and symbolic). Local people perceive values based on their everyday interrelations and interdependence.

The IPBES conceptual framework (Díaz et al., 2015) focuses on multiple values of biodiversity and ecosystem services by dividing it into three main categories namely nature, nature's benefits to people and good quality of life. Multiple valuations can identify a set of decisions, embedded in an institutional context. This can create a good quality of life by ensuring the smooth flow of nature's benefits to the people. However, there have been many formulations and definitions of human well-being (Alkire, 2002). It includes basic materialistic provision for a decent life, freedom, health, social security and relations that contribute to physical, social, psychological and spiritual well-being. These factors impact each other, either positively or negatively. A change in one often brings out changes in the other like a ripple effect. This welfare-based framework tells us little about the well-being of nature.

Well-Being and Ecosystem Services: Interconnectedness and Cooperation Between Human and Nature

The following framework demonstrates the interconnectedness between humans and nature, along with well-being resulting from the benefits received from ES that is subjected to production, ecological and political

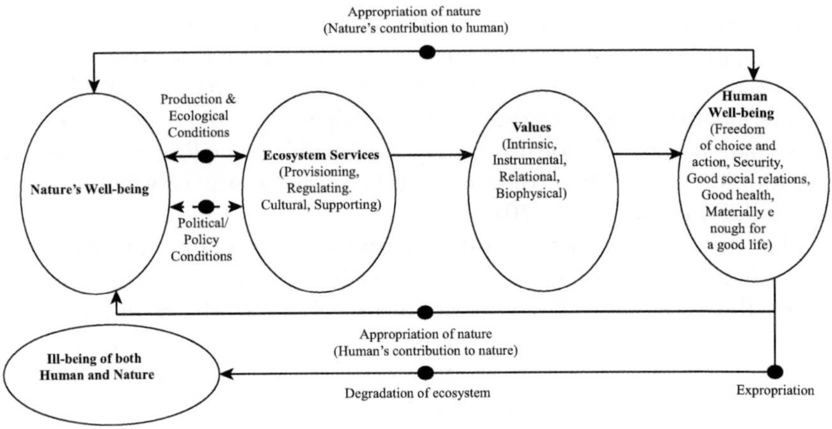

Fig. 6.1 Ecosystem services and well-being (*Source* Adapted from Titumir & Paran, 2022b)

conditions (Fig. 6.1). It reveals that different aspects of ES keep thriving if nature's well-being is maintained. When the drivers of change (production, ecological and political conditions) affect the ecosystem positively, ecosystem service derived from nature's well-being is passed on as a contribution to human well-being in various forms. Similarly, if the drivers negatively affect the ecosystem, its impact is negative on the well-being of both nature and humans.

The appropriation of nature ensures a harmonious relationship between humans and nature which in turn has a positive impact on the quality of both nature and human life. Identifying the direct (production and ecological conditions) and indirect drivers (policy/political conditions), this framework depicts that transformative changes occur if each other's interdependency, access and benefit-sharing and well-being is valued, apart from treating the environment as economic goods alone (Figs. 6.1 and 6.2). Exploring human and nature's interconnected relationship also categorises nature's contribution to human (provisioning, regulating and cultural) as well as human's contribution to nature (anthropogenic assets and drivers). All these are interlinked towards maintaining the well-being and sustainability of both. The drivers are understood by taking into account carbon emission, pollution, over-extraction, degradation and exclusion, harvesting, fishing, reforestation, innovative use of resources, climate change, weather patterns, natural disaster and hazards, etc., within SEPLS (Fig. 6.2).

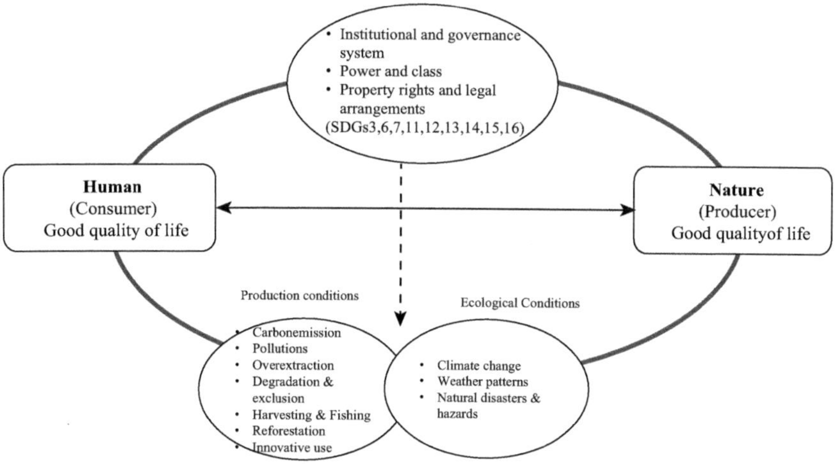

Fig. 6.2 Transformative pathways for human-nature relationships (*Source* Adapted from Titumir & Paran, 2022b)

The Current State of ES: TRUs' Perspective

The TRUs are dependent on the forest for goods and services which contributes towards their well-being (see Titumir & Paran, 2022a, 2022b, 2022c). At the same time, customary and traditional rules and practices of TRUs help to conserve the ecosystem (Titumir et al., 2019, 2020). The mutuality and cooperation between the forest and the TRUs have been identified as a significant precondition towards each other's well-being.

Four FGDs were conducted among four different groups of TRUs of *Koyra Bonojibi Samiti*. These four groups (*Bawali*, *Moual*, *Jele* and crab collectors) were categorised by their respective goods and services. These goods and services have numerous values which contribute to well-being. Each of the groups developed a resource map depicting the distribution of respective resources in the forest. The woodcutters (*Bawali*) identified and drew the resource map on flora, fishermen (*Jele*) categorised and drew the resource map on fish, honey collectors (*Moual*) depicted the distribution of beehive (*mouchak*) in the forest about honey and wax and crab collectors told about the distribution of crabs. The distribution of resources in the case study area is drawn on the resource maps (Figs. 6.3, 6.4, 6.5, and 6.6).

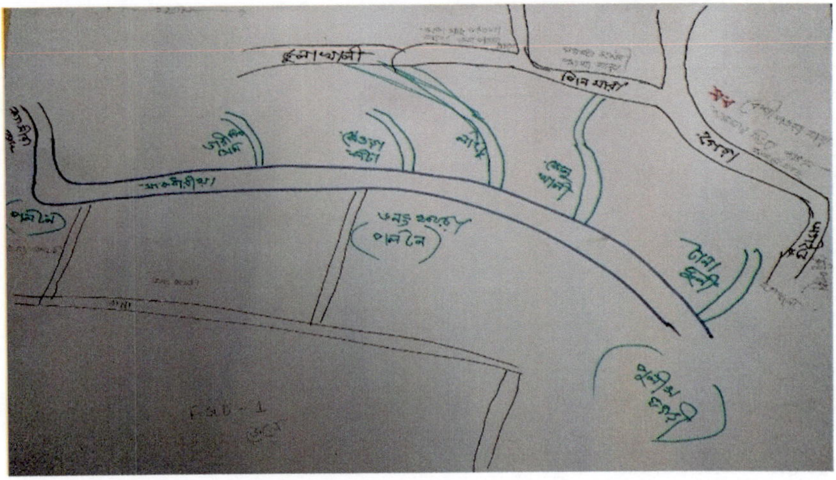

Fig. 6.3 Distribution of fish drawn by fishermen (*Jele*) (*Source* Unnayan Onneshan)

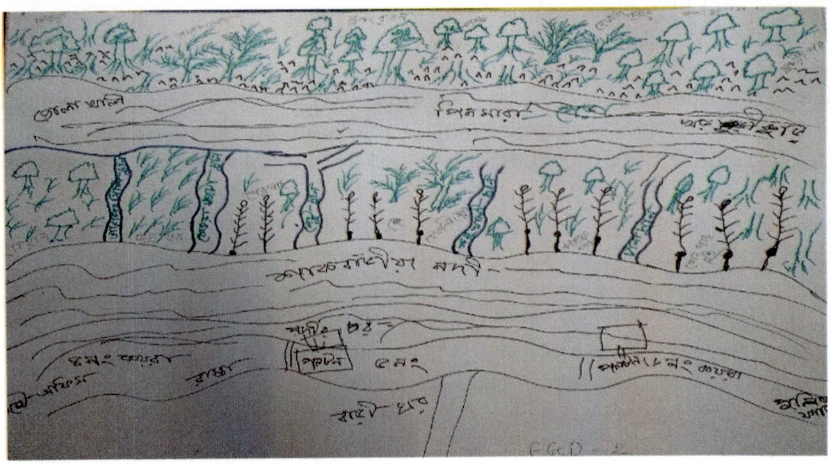

Fig. 6.4 Distribution of flora drawn by woodcutters (*Bawalis*) (*Source* Unnayan Onneshan)

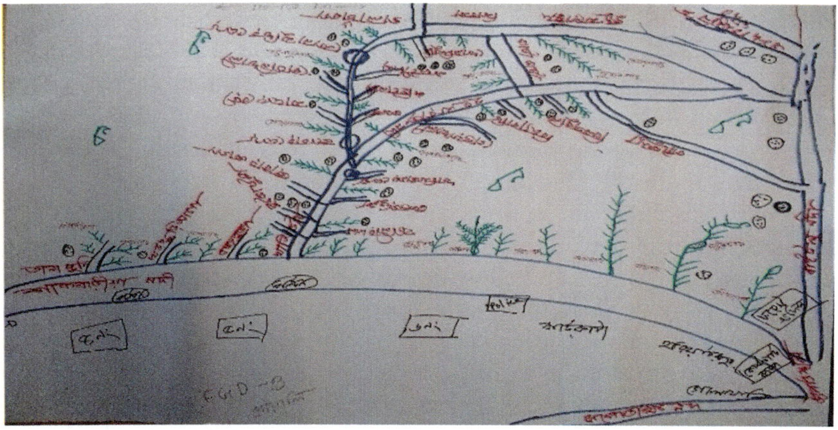

Fig. 6.5 Distribution of beehives drawn by the honey and wax collectors (*Mouals*) (*Source* Unnayan Onneshan)

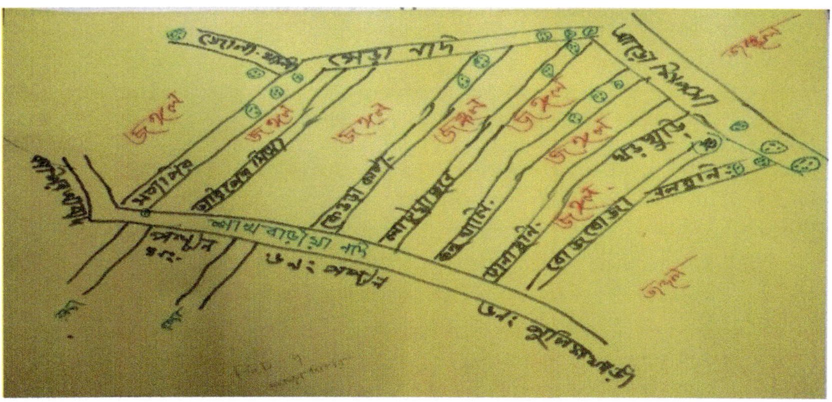

Fig. 6.6 Distribution of crabs drawn by crab collectors (*Source* Unnayan Onneshan)

Fishermen (Jele)

This group talked about various fish which they collect from the water-bodies in the forest. Their fish catch consists of *java*, *dadnee*, *shrimp* (*bagda*, *chai* and *chaka*), *vetki*, *paissha*, *baila*, *chanda*, etc. Most of these fish are sold in the market and at times some of these fish are consumed by the fishermen and their family members. They added that some fish are used for medicinal purposes (for example *java fish*) which fetches a high price for the catch at the market. All the six members from the *jele* (fishermen) community who participated in the FDG said that the volume of fish catch was steadily declining. They attributed it to the use of pernicious nets like *bishal jal* and the poisoning of the water bodies as key factors behind the depletion. Forest officials, police and other governing bodies often take bribes and allow powerful groups to fish in the area and these people adopt unsustainable extraction practices. Punitive action is hardly employed against them which is supposed to be the task of the forest officials. In addition to it the flow of water from various tributaries in this delta region on which the mangrove forest thrives is declining that causes serious damage to different species of fish in the water bodies of the forest (see drivers of changes in Table 6.2). Figure 6.3 shows the distribution of fish in different water bodies in the area of study. The fishermen demonstrated the availability of fish species and the current state of reproduction in different canals and rivers through a resource mapping process.

Woodcutters (Bawali)

Collection and cutting of wood within the premises of the forest are banned. The local *Bawali* (woodcutter), however, collect *golpata* and different species of creepers used for medicinal purposes and serves as fuel. They categorised various floral species like *bain*, *keora*, *goran*, *geowa*, *sundari*, *pashur*, *dhundhul*, *kakra*, *khailsa*, *hental*, *bonshosha*, *bonlebu*, *choitboroi*, *gab*, *onra*, *bhuikumra*, etc., as available in the forest. However, all these species are not found in the same area. Distribution varies based on the salinity of water in the forest. They said that powerful syndicates are often involved in cutting timber illegally, even though local *bawali* are not allowed to do so. The decline in the forest area is happening due to this kind of illegal extraction, as well as frequent devastation caused by cyclones. Natural disasters like cyclones have had a negative impact on the forest cover to a great extent. In addition, when a big tree is cut down, it destroys all other small trees and plants surrounding it. The distribution of floral species is depicted in a resource map drawn by *Bawali* during the FGD (Fig. 6.4).

Honey and Wax Collectors (Moual)

Honey and wax are collected from the forest. All six participants from this group said during the FGD that the availability of honeycombs is better in the dense forest. Hence, it is obvious that if the number of trees declines, the number of honeycombs will also decline. *Moual* follow some traditional rules while collecting honey. They collect honey and wax during April, May and June. They cut a specific section (about two-thirds) of the honeycomb and leave the rest for reproduction. This process ensures that no young bees are killed during honey collection. In addition, beehives are squeezed by hand. They never use metal tools that may be dangerous for the bees. They produce smoke using dry leaves but never put the beehive on fire. Meanwhile, the outsiders who do not know the traditional practices and rules often end up killing young bees. Moreover, a group of people also cut the entire beehives in order to collect wax that kills young bees and destroys future reproduction. During FGD the *moual* came up with a solution that can stall this unsustainable practice employed by outsiders. Based on their experience the *moual* have discovered that a beehive that produces 40 kg honey will have 2–2.5 kg of wax when the hive is not cut completely. Therefore, if a honey collector collects more than 2–2.5 kg of wax against a collection of 40 kg honey, it implies that the hive has been completely annihilated. If this is taken into consideration then people employing unsustainable means to collect honey could be easily identified. However, implementing such calculations and identifying the miscreants depends on the effectiveness of administrative bodies, who are often hand in gloves with people involved in deploying unfair means of resource extraction from the forest. The distribution of beehives is drawn by the *moual*, showing that more honey is found inside the deep forest (Fig. 6.5).

Crab Collectors

Crab collectors who are not the TRUs, use harmful nets like *bishal jal* and poison the water. This kills eggs, small crabs as well as the breeding ground of crabs. Such unsustainable practices are now posing major threats to crabs and other aquatic resources, even though crab collection has not declined significantly. TRUs say that more crabs are caught in muddy water. If there is an increased flow of water in the rivers and canals, water becomes muddy, thus increasing the number of crabs. The flow of water, however, has decreased due to the presence of dams and barrages in the upstream rivers. As the number of crab collectors is increasing gradually, the number of crabs in the waterways is decreasing.

Landscape Value Mapping of Woodcutters

Valuable wood or trees are not widespread throughout the forest. These trees can be seen in certain areas. Areas, where big and high-priced trees or timber are available, are more valuable and more trees are extracted from those areas. The dark red part here indicates the most valuable part of the forest where big and precious trees are in abundance. The medium red part means that high-priced wood is found here but less compared to the dark red part. And in the light red part also some valuable trees or wood can be found. Similarly, the higher the red part, the greater the amount of timber extracted in those areas (Fig. 6.7).

Landscape value mapping of woodcutters ascertains that the number of big trees has declined significantly in the last two decades (2000–2020). According to the data provided by the *Bawali* during *PPGIS*, it is evident

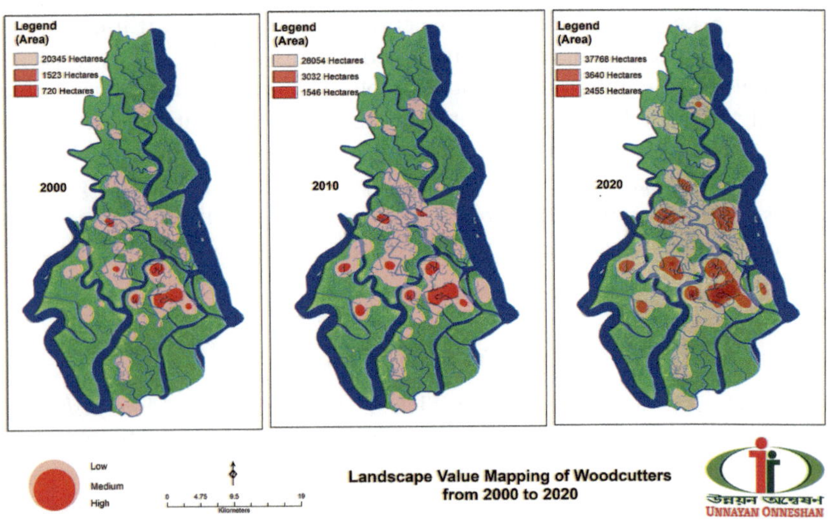

Fig. 6.7 Landscape value mapping of woodcutters (*Source* Unnayan Onneshan, 2020)

from the map of landscape valuation that the value of different areas of the forest has increased from time to time based on the availability of big and precious trees in the forest. In the past two decades, the value of the area where precious timber or trees are found has increased a lot. As a result of which, the extraction of valuable trees from these areas has also increased, resulting in decline of big trees in the forest (Fig. 6.7).

The average growth rate of the dark red part in two decades is 86.75%. In 2000, this area was 720 hectares. In a decade, it increased by 114.72% to 1,546 hectares in 2010. In the following decade, the dark red area increased by 58.79% and in 2020 it stood at 2,455 hectares. Similarly, the medium red and the light red parts have increased. In two decades, the average growth rate of the medium red part is 59.56% and the average growth rate of the light red part is 36.25% (Fig. 6.7).

In 2000, the medium red part was 1,523 hectares. In one decade, it increased by 99.08% and in 2010 it stood at 3,032 hectares. In other words, values have almost doubled in such areas. In 2010–2020, the medium red part increased by 20.05% and stood at 3,640 hectares in 2020.

On the other hand, in 2000, the amount of light red part was 20,345 hectares. In 2010, the area increased by 37.89% to 28,054 hectares. In the following decade, the light red part has increased by 34.62%. In 2020, the size of such part stands at 37,766 hectares (Fig. 6.7). The massive increase in the value of different areas of the forest on the basis of precious timber indicates a massive increase in extraction of trees in two decades. As a result, there has been a significant reduction in forest coverage during this time (see Chapter 2, Fig. 2.4). Similar findings came out from the resource map drawing by TRUs (Fig. 6.4).

Ecological Value Mapping of Honey and Wax Collectors (Moual)

Honey and wax are not available in every part of the forest. Bees usually nest in dense forests where there are larger trees. Areas, where honey and wax are available, are more valuable. So, more honey and beeswax are extracted from those areas. The dark red part here refers to the most valuable part of the forest based on the availability of honey and beeswax. The medium red part means honey and wax are also found here but less than the dark red part. The light red part means that some beehives can be seen here too. The higher the red part, the higher the amount of honey and wax extracted in those areas (Fig. 6.8).

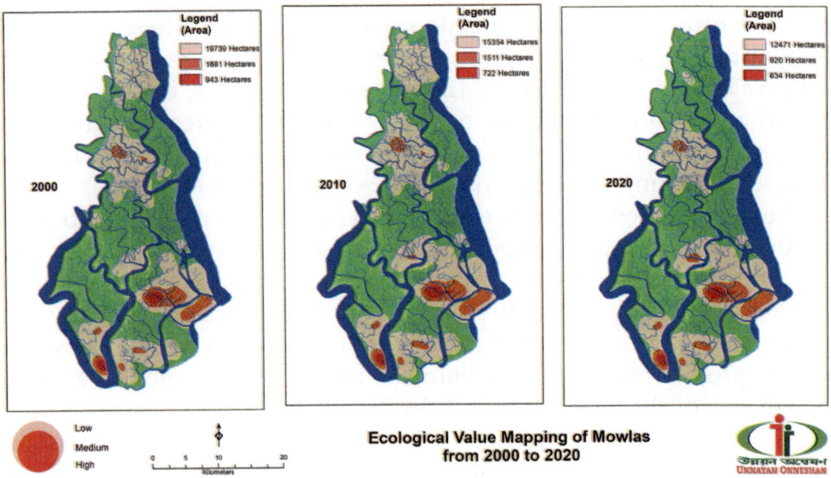

Fig. 6.8 Ecological value mapping of honey and wax collectors (*Source* Unnayan Onneshan, 2020)

In the last two decades (2000–2020), the number of beehives has declined in the forest. According to the data provided by the honey and wax collectors, it appears from the map that in two decades (2000–2020) the value of different areas of the forest has declined significantly based on the availability of honey and beeswax. The value of different regions of the forest decreased as the availability of honey and wax declined (Fig. 6.8).

In two decades, the dark red part has decreased by an average of 17.80%. In 2000, the area under dark red was 943 hectares, which decreased by 23.43% in ten years and stood at 722 hectares in 2010. In the following ten years, it decreased by 12.18% and stood at 634 hectares in 2020.

Similarly, the medium red part and the light red part has decreased by an average of 29.39 and 20.49%, respectively, in two decades. In 2000, the medium red area was 1,881 hectares, which decreased by 19.7% in a decade to 1,511 hectares in 2010. From 2010 to 2020, the medium red part further decreased (39.11%) to 920 hectares in 2020. On the other hand, the light red part in 2000 was 19,739 hectares, which decreased by 22.21% in ten years to 15,354 hectares in 2010. In the following decade, with a further reduction of 18.77% the light red part stood at 12,471 hectares in 2020 (Fig. 6.8).

Findings from ecological value mapping of honey and wax suggest that the availability of these resources is declining gradually. As a result, the value of different parts of the forest in terms of honey and wax is decreasing in this region. A similar outcome came from the FGD (Fig. 6.5).

Ecological Value Mapping of Fishermen and Crab Collectors

The ecological value mapping of fishermen and crab collectors also depicts the degradation of fishing areas and crab collecting areas in the water bodies of the forest, respectively (see Figs. 3.1 and 3.2 in Chapter 3). The value of different parts of the rivers, canals and creeks has significantly decreased in two decades that indicates that over-extraction of fishes and crabs has been going on in those areas leading to less availability of aquatic resources. These findings also corroborate with the FGDs conducted among fishermen and crab collectors, respectively (Figs. 6.3 and 6.6).

Ecosystem Services and Multiple Values

Most of the valuation methods adopt a market-centric approach and estimate only direct or indirect use of value (Table 6.1). These value approaches unfortunately cannot capture all the intrinsic, instrumental, biophysical, relational values of natural resources (Titumir et al., 2019). All the services from the mangrove ecosystem in the form of provisioning, regulating, supporting and cultural cannot be estimated in a market-based monetary approach. Apart from fisheries, forestry, erosion control, carbon sequestration and tourism, plenty of services are obtained from the mangrove ecosystem (Table 6.2).

TRUs' Perception

This study adopted Likert scale in order to rank the degrees of importance of ES by the TRUs. Total 11 types of values and the importance of ES have been explored using the five-point scale. The degrees of importance of multiple values are ranked as highly important, important, somewhat important, not important and do not know. The number of respondents for exploring each type of value was 115 (Table 6.2).

Table 6.1 Valuation of mangroves ecosystem services in different studies

Valuation method	*Direct use value*	*Indirect use value*	*Author*
Market price	Local uses: fruits, cigarette wrappers and nipa thatch for roofing, on site fisheries	Off-site fisheries	Christensen (1982)
Market price/shadow price	On-site fisheries, forestry, agriculture and aquaculture	Off-site fisheries, nutrient	Lal (1990)
Market price	On-site fisheries, forestry	Off-site fisheries, tourist industry	Bennet and Reynolds (1993)
Market price/shadow price	Local use, forestry, on site fishery	Erosion control, offsite fishery, biodiversity	Ruitenbeek (1992)
Market price	Local use, forestry, on site fishery	Off-site fisheries	Gammage (1994)
Market price	Forestry	–	Gilbert AJ and Janssen (1998)
Cost and benefit analysis	Forestry, fisheries	Erosion control, carbon sequestration	Sathirathai (1998)

Source Adapted and modified from Vo et al. (2012)

Out of all the respondents, 70% believe that the natural beauty of the forest is highly important, while the rest 30% ranked it as important. Of the respondents, 52% ranked the historical values as important, while 41% said it is somewhat important. Only 2% of the respondents said that they do not know about the historical value of the forest. In terms of education and knowledge, 26% identified it as highly important, 32% identified it as important and 42% identified it as somewhat important. The value of tourism, recreation and essentiality beneficial for future generations, cultural and local heritage has been identified as significant to TRUs. Most respondents claim that these are important or somewhat important to them. Regarding the religious and spiritual importance, 64% explained it to be highly important, while 25% explained it to be important.

On the other hand, 77% of the respondents believe that the forest is highly important to them for their lives and livelihoods. Besides, 25% identified the forest as highly important for keeping natural balance, while 24 and 50% identified the importance of keeping natural balance as important and somewhat important, respectively. Most of the respondents also

Table 6.2 Degrees of importance of multiple values to the TRUs

Value/importance	*No. of respondent*	*Highly important (%)*	*Important (%)*	*Somewhat important (%)*	*Not important (%)*	*Do not know (%)*
Natural beauty	115	70	30	0	0	0
Historical value	115	5	52	41	0	2
Education and knowledge	115	26	32	42	0	0
Tourism and recreation	115	17	43	41	0	0
Beneficial for future generation	115	6	50	44	0	0
Local and cultural heritage and identity	115	17	39	44	0	0
Religious and spiritual	115	64	25	9	2	0
Life and livelihoods	115	77	17	7	0	0
Keeping natural balance	115	25	24	50	0	0
Combating climate change	115	15	34	51	0	0
Medicinal resources	115	23	23	51	1	1

Source Authors' calculations

view the value of the forest as highly important (15%), important (34%) or somewhat important (51%) in coping with climate change. The importance as a source of medicinal resources has been found to be significant too (Table 6.2).

Ecosystem Services and Value Types

The ES of the forest contributing to well-being can be categorised on the degrees of importance of the multiple values reflected in the survey and the state of ecosystem services as reflected in the FGDs. The types of ecosystem services are understood according to the millennium assessment (2005a) that has been categorised into multiple values following the IPBES valuation approach (2015). The usage or importance of certain values, that is, regulating and supporting services are retrieved from scientific literature (Table 6.3).

The provisioning services (foods, woods, fuel and water) of natural resources recommend the instrumental value. The supporting services (nutrient cycle, soil formation and primary production) are the underlying and biophysical values of nature. The regulating services (regulation of climate and flood, disease prevention and water purification) also record the instrumental value. On the other hand, cultural services are mainly related to relational values. The different items/usage of resources and services contributing to the well-being and their corresponding values are depicted in Table 6.3.

The forest provides livelihoods, enjoyment and recreation. It also serves as the protecting belt of the coastal region. The TRUs associate this forest as their nourisher on which their existence depends. The values (Table 6.3) contribute to the well-being of the TRUs. Many factors that affect the well-being of both the forest and humans. When these factors are altered, the provision of well-being is also altered.

Drivers of Changes

The direct and indirect drivers affect the well-being of the forest and the TRUs. The main problems identified by the TRUs during FGDs include climate change, over-extraction of resources, unsustainable practices such as poisoning, rent-seeking behaviour of forest departments and institutional fragility (Table 6.4). The anthropogenic pressure is damaging the forest owing to weak institutional and governance systems.

The changes in the well-being of the forest due to the production, ecological and political conditions are negatively affecting the well-being of the human being. The Likert scale analysis depicts the impacts of climate changes affecting the forest, biodiversity along with the life and livelihoods of the TRUs. The analysis shows that the salinity intrusion is increasing gradually and causing serious damage to life, livelihoods and biodiversity. The resources are depleting due to over-extraction and changing climate variables as well. All these impacts are adversely affecting the well-being of the TRUs.

Table 6.3 Ecosystem services and types of values

Ecosystem services	*Item/usage contributing to well-being*	*Types of value*
Provisioning services		
Foods	Fish, shrimp, prawns, shells, honey, salt, crabs, fruits	Instrumental value
Woods	Furniture, boats, pillars of house	Instrumental value
Fuel	For household cooking	Instrumental value
Water	Transportation, shrimp and crab cultivation	Instrumental value
Supporting services		
Nutrient cycle	Mineralisation of nitrogen and phosphorus by fish through excretion	Intrinsic value
Soil formation	Forest (roots of trees) prevents soil erosion	Intrinsic value
Primary production	Removing toxic elements like SO_2, CO_2 to reduce air pollution and reproduction of various species of flora and fauna, raw materials (algae, sand, seaweed)	Intrinsic value, Biophysical value
Regulating services		
Climate regulation	Natural wall against climate change such as storms, cyclones and salinity intrusion	Instrumental value
Flood regulation	Protection from storms, floods, tsunami, etc	Instrumental value
Disease prevention	Medicinal usage of different floral and faunal species such as Cortex of Poshur (*X. mekongensis*), Hargoza (*Acanthus ilicifolius*), fruits of Sundori (*Heritiera fomes*)	Instrumental value
Water purification	Prevents salinity, rivers and waterways work for movement and storage of water	Instrumental value
Cultural services		
Aesthetic	Seascape, landscape, abundance of beautiful biodiversity, etc	Relational
Spiritual	Certain beliefs such as '*Bonobibi*' and religious rituals	Relational
Educational	Knowledge creation and research scope	Relational and Instrumental value
Recreational	Nature watching, sailing, recreational fishing, etc.	Relational and Instrumental value

Source Adapted from Titumir & Paran, 2022b

Table 6.4 Drivers of changes in the Sundarbans

Drivers	*Remarks*
Direct drivers	
Production conditions	
Carbon emissions	Developmental projects near the forest such as *Rampal* coal-based power plant and industries emit CO_2 and other toxic gases that negatively affect the ecosystem of the Sundarbans
Pollutions	Air pollution due to emissions of toxic gases, water pollution due to salinity intrusion along with industrial wastage and soil pollution due to increased salinity from sea level rise and shrimp farming negatively affecting the ecosystem
Extraction of resources	Unsustainable and over-extraction of resources degrades biodiversity and ecosystem
Degradation and exclusion	Rent-seeking behaviour induces over-extraction which leads to the degradation of resources where powerless and poor section gets excluded
Harvesting	Unsustainable harvesting like shrimp farming near the forest harms the ecosystem
Fishing	Use of Pernicious nets, poisons to catch fish and over-extraction leads to resource degradation
Reforestation	Community afforestation near the rivers and embankments benefit the ecosystem
Innovative use of resources	Customary sustainable practices and innovative use of resources, such as CMAASC help conserve the natural forest creating alternative livelihood options
Ecological conditions	
Climate change	Climate change negatively affects the ecosystem, frequent cyclones, floods, increased salinity and sea-level rise
Weather patterns	Change in weather patterns due to climate change negatively affects the ecosystem
Natural disasters and hazards	Frequently hit the ecosystem causing serious harm
Indirect drivers	
Political/policy level conditions	
Institutions	Institutions are the rules of the games, extractive institutions are causing over-extraction of natural resources
Governance	Weak and corrupted governance indulges in rent-seeking practices and fails to conserve the forest

(continued)

Table 6.4 (continued)

Drivers	*Remarks*
Power and class	Powerful syndicate uses rents to have illegal extraction of resources
Property rights	TRUs are denied their rights which causes over-extraction of the resources
Legal arrangements	Low enforcement of existing laws which barely protects TRUs' perspectives

Source Adapted from Titumir & Paran, 2022b

Well-Being Perception Among the TRUs

A Likert scale analysis is used to depict the perception of the TRUs regarding well-being. A total of eight questions were asked during this survey. There were randomly selected 115 respondents who are TRUs and members of the Koyra Forest People Development Cooperative. The answers were explored using a four-point scale (highly satisfied, satisfied, somewhat satisfied and not satisfied at all) asking how satisfied they were in terms of life and livelihoods, security, health, freedom of choice and social relations.

The first question asked how satisfied the respondents were with their life. Of all the respondents, 43% replied that they were satisfied with their life, while 32% replied that they were highly satisfied. On the other hand, 25% said that they were somewhat satisfied. The second question was about the satisfaction of their lifestyle and the success of their activities. The majority of the respondents (60%) replied that they were somewhat satisfied and only 2% replied that they were not satisfied at all. The third question was how happy they were with their life. The majority (56%) in this case replied that they were somewhat satisfied with happiness in life, while others responded to being satisfied (23%) or highly satisfied (19%). The fourth question inquired the satisfaction level about the security of life and livelihoods of the TRUs. The majority of the respondents (69%) argued that they were somewhat satisfied. The fifth question explored the level of satisfaction in terms of health. In this case, again the majority of the respondents (63%) replied that they were somewhat satisfied. The sixth one inquired the satisfaction regarding their social relations. Being satisfied, 35% of the respondent implied that they had good social relations. The majority of the respondents (60%) were somewhat satisfied.

The seventh question asked about the level of satisfaction, regarding the ability to meet the basic needs and demands of life. Again, the majority of the respondents (66%) said that they were somewhat satisfied. The eighth question asked about the satisfaction level regarding freedom of choice and freedom of profession. The majority of the respondents (57%) believe that they were highly satisfied with this, while 23% believe that they were satisfied. In this case, 6% of the respondents replied that they were not satisfied at all.

The Likert scale analysis implied that, in most of the cases, the majority of the respondents were somewhat satisfied. The majority of the TRUs perceive that they were not fully satisfied with life, livelihoods, security, health, social relations and freedom of choice (Fig. 6.9).

Corroborative literature also supports the findings from the Likert scale analysis adopting the perception of TRUs regarding the context of well-being. It is observed that species such as *Heritiera* and *Ceriops* are on the brink of extinction in the future, increasing the amount of *Excoecaria* and *Bruguiera* which is abundant at present. This kind of ecosystem will be unable to provide services, like high carbon sequestration rate (Nandy, 2001), good timber and a few traditional uses of

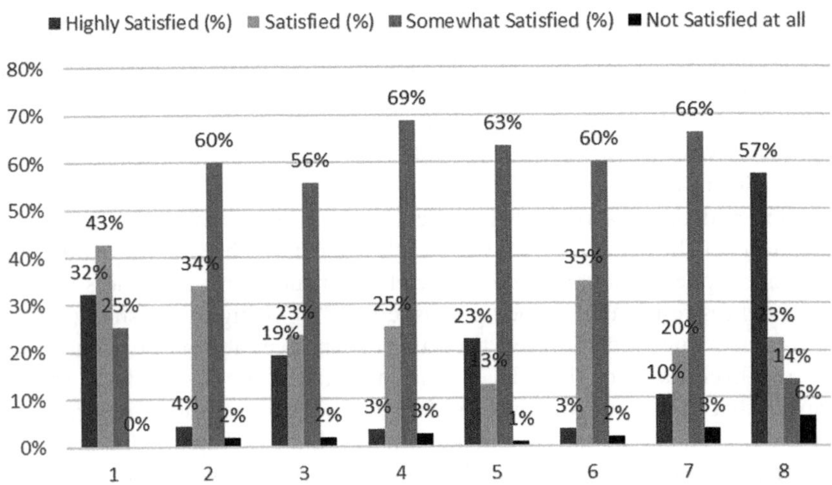

Fig. 6.9 Well-being perception of TRUs (*Source* Adapted from Titumir & Paran, 2022b)

medicine that are regarded as the specialty of *Heritiera* and *Ceriops*. There is also a possibility of a decrease of services like storm and shoreline protection, and provision of primary production. On the counter position, increasing abundance of *Excoecaria* and *Bruguiera* and their products would be utilised better than it is being now for furniture and medicinal use (Konoshima et al., 2001). Overall, since the forest extent is forecasted to fall short by the end of 2100, the magnitude of ecosystem services is expected to reduce. This is especially from those species that are less tolerant to salt (Mukhopadhyay et al., 2018).

Conclusions

This chapter deals with the multiple values of ecosystem services and attempts to put value to them by attaching them to human well-being derived from the forest and the benefits we receive from it. The benefits from the ecosystems appear in the form of various goods and services, known as ecosystem services that help to make human life both possible and worth living. It was also revealed that the types of ecosystem services keep thriving if nature's well-being is maintained. If the drivers of change such as the production, ecological and political conditions affect the ecosystem positively, then the services of the ecosystem derive from nature's well-being and contribute towards human well-being in many ways; and vice versa. The current state of ecosystem services from the TRU's perspective also revealed that those who were involved in collecting fish said the amount of fish collection was steadily declining due to pernicious nets and poisoning of water bodies that was done by the powerful groups. Similarly, the honey collectors also revealed how outsiders ruin the beehives and kill young bees through harmful honey-collecting practices. The ecosystem services that are discussed were also classified into many categories; some of which included services such as provisioning, regulating, supporting, etc.

The analysis demonstrates that a market-centric approach cannot reflect the multiple values of the ecosystem services. TRUs value the Sundarbans as their life. The values of the ecosystem services are enormous in contributing to the survival and existence of the people. The well-being of the people relies on the well-being of the forest. Climate change and its rising impacts in the form of cyclones, salinity intrusion, sea-level rise is causing damage to the forest and its people (see Chapter 8 for more). Although the forest is protecting people and the region from climate change and its negative impacts through various regulating services, it has to bear serious damages too. The over-extraction of resources in addition to pollution from various anthropogenic variables are resulting in massive depletion of biodiversity and habitats of the forest. The fragile institutions and ineffective governance systems are contributing to the destruction of the ecosystem in the forest.

To analyse the well-being perception among the TRUs, a survey was carried out and a Likert scale was used for evaluation. The questions that were asked were regarding how satisfied they were in terms of life and livelihoods, security, health, freedom of choice and social relations. The analysis implied that, in most of the cases, the majority of the respondents were somewhat satisfied. In addition to this, corroborative literature also supports the findings from the Likert scale analysis adopting the perception of TRUs regarding the context of well-being.

As a result, the well-being provision of the human being—TRUs is declining. In terms of life and livelihoods, health, security, social relations and their freedom of choice, people are becoming less satisfied day by day. It can be seen that the traditional rules and practices can significantly help in maintaining the natural balance of the forest (see Chapter 4, 5 and 12). This can occur if it is adopted through a formal governance system. Since the well-being of humans and nature is interconnected, the appropriation of natural resources by maintaining a balance in the ecosystem becomes crucial.

Acknowledgements Many insights of the chapter are drawn from 'Ecosystem services and well-being in the Sundarbans of Bangladesh: A multiple evidence base trajectory', to be published by Springer Nature, Singapore in *Assessing, mapping and modelling of mangrove ecosystem services*, eds. R. Dasgupta, S. Hashimoto & O. Saito, Springer Sustainability Science Book Series.

References

Alkire, S. (2002). Dimensions of human development. *World Development, 30*(2), 181–205.

Banzhaf, S., & Boyd, J. (2007). What are ecosystem services? The need for standardized environmental accounting units. *Ecological Economics, 63*(2–3), 616–626.

Bengtsson, J. (1997). Which species? What kind of diversity? Which ecosystem function? Some problems in studies of relations between biodiversity and ecosystem function. *Applied Soil Ecology, 10*, 191–199.

Costanza, R., D'Arge, R., de Groot, R., Farberk, S., Grasso, M., Hannon, B., Limburg, K., Naeem, S., O'Neill, R. V., Paruelo, J., Raskin, R. G., Sutton, P., & van den Belt, M. (1997). The value of the world's ecosystem services and natural capital. *Nature, 387*, 253–260.

Daily, G. (1997). *Nature's services: Societal dependence on natural ecosystems.* Island Press.

Daily, G. C., Alexander, S., Ehrlich, P. R., Goulder, L., Lubchenco, J., Matson, P. A., Mooney, H. A., Postel, S., Schneider, S. H., Tilman, D., & Woodwell, G. M. (1997). Ecosystem service: Benefits supplied to human societies by natural ecosystems. *Issues in Ecology, 1*(2), 1–18.

De Groot, R. S., Wilson, M. A., & Boumans, R. M. J. (2002). A typology for the classification, description and valuation of ecosystem functions, goods and services. *Ecological Economics, 41*, 393–408.

Díaz, S., Fargione, J., Chapin, F. S. III, & Tilman, D. (2006). Biodiversity loss threatens human well-being. *PLoS Biol, 4*(8), e277. https://doi.org/10.1371/journal.pbio.0040277

Díaz, S., Demissew, S., Carabias, J., Joly, C., Lonsdale, M., Ash, N., et al. (2015). The IPBES conceptual framework—Connecting nature and people. *Elsevier, 14*, 1–16.

González-Jiménez, D., Berghöfer, U., Berghöfer, A., Heubach, K., Kosmus, M., & Bertrab-Tamm, A. V. (2018). *Beyond measurements: Multiple values of nature and their diverse conceptualization.* Intergovernmental Science-Policy Platform on Biodiversity and Ecosystem Services (IPBES), Bonn, Germany.

King, D. M., Wainger, L. A., Bartoldus, C. C., & Wakeley, J. S. (2000). *Expanding wetland assessment procedures: Linking indices of wetland function with services and values.* Wetland Research Program, Washington, DC, USA.

Konoshima, T., Konishi, T., Takasaki, M., Yamazoe, K., & Tokuda, H. (2001). Anti-tumor-promoting activity of the Diterpene from Excoecaria agallocha. *Biological &/and Pharmaceutical Bulletin, 24*(12), 1440–1442.

Layke, C., Mapendembe, A., Brown, C., Walpole, M., & Winn, J. (2012). Indicators from the global and sub-global millennium ecosystem assessments: An analysis and next steps. *Ecological Indicators, 17*, 77–87. https://doi.org/10.1016/j.ecolind.2011.04.025

Millennium Ecosystem Assessment. (2005a). *Ecosystems and human well-being: Current state and trends.* Island Press.

Millennium Ecosystem Assessment. (2005b). *Ecosystems and human well-being: A framework for assessment.* Island Press.

Mukhopadhyay, A., Payo, A., Chanda, A., Ghosh, T., Chowdhury, S. M., & Hazra, S. (2018). Dynamics of the Sundarbans mangroves in Bangladesh under climate change. In R. Nicholls, C. Hutton, W. Adger, S. Hanson, M. Rahman, & M. Salehin (Eds.), *Ecosystem services for well-being in deltas*. Palgrave Macmillan.

Nandy, G. J. (2001). Relation between rate of photosynthesis and leaf temperature during summer at specific PAR values. ~, air temp. *Plant Biol, 44*(s4).

Titumir, R. A. M., Afrin, T., & Islam, M. S. (2020). Traditional knowledge, institutions and human sociality in sustainable use and conservation of biodiversity of the Sundarbans of Bangladesh. In O. Saito, S. M. Subramanian, H. Hashimoto, & K. Takeuchi (Eds.), *Managing socio-ecological production landscapes and seascapes for sustainable communities in Asia: Mapping and navigating stakeholders, policy and action* (pp. 67–92). Science for Sustainable Societies. https://doi.org/10.1007/978-981-15-1133-2_5

Titumir, R. A. M., & Paran, M. S. (2022a). Human-nature cooperation for well-being: Community understanding on one health approach in the COVID-19 era in the Sundarbans. In M. Nishi, S. M. Subramanian, H. Gupta (Eds.), *Biodiversity-health-sustainability nexus in socio-ecological production landscapes and seascapes (SEPLS)*. Springer Nature, Singapore.

Titumir, R. A. M., & Paran, M. S. (2022b). Ecosystem services and well-being in the Sundarbans of Bangladesh: A multiple evidence base trajectory. In R. Dasgupta, S. Hashimoto, & O. Saito (Eds.), *Assessing, mapping and modelling of mangrove ecosystem services*. Springer Sustainability Science Book Series (Forthcoming).

Titumir, R. A. M., & Paran, M. S. (2022c). Human-nature relationship for post 2020 biodiversity framework: An exploration into traditional knowledge, customary sustainable use and community-based innovation in the Sundarbans. In R. A. M. Titumir (Ed.), *Fifty years of nation building: Political economy of Bangladesh's development*. Centennial Book Series Project of University of Dhaka. Department of Development Studies, University of Dhaka (Forthcoming).

Titumir, R. A. M., Paran, M. S., & Pasha, M. W. (2019). The Sundarbans is our mind: An exploration into multiple values of nature in conversation with traditional resource users. In UNU-IAS & IGES (Eds.), *Understanding the multiple values associated with sustainable use in socio-ecological production landscapes and seascapes (SEPLS)* (Vol. 5, pp. 97–117). Satoyama Initiative Thematic Review. United Nations University Institute for the Advanced Study of Sustainability.

Unnayan Onneshan. (2020). *Biodiversity, climate change and traditional resource users in the Sundarbans: An exploration through public participation geographic information system (PPGIS)* (Unpublished Research Report). Unnayan Onneshan, Dhaka, Bangladesh.

van Oudenhoven, A. P. E., Petz, K., Alkemade, R., Hein, L., & de Groot, R. S. (2012). Framework for systematic indicator selection to assess effects of land management on ecosystem services. *Ecological Indicators, 21*, 110–122. https://doi.org/10.1016/j.ecolind.2012.01.012

Vo, T., Kuenzer, C., Minh, V., Moder, F., & Oppelt, N. (2012). Review of valuation methods for mangrove ecosystem services. *Ecological Indicators, 23*, 431–446.

Wallace, K. J. (2007). Classification of ecosystem services: Problems and solutions. *Biological Conservation, 139*(3–4), 235–246.

CHAPTER 7

Multiple Values of Nature and Transformational Pathways

Rashed Al Mahmud Titumir, Md. Shah Paran, Mostafa Walid Pasha, and Mohammad Saeed Islam

Introduction

This chapter aims to assess the value of multiple resources in the Sundarbans and attempts towards developing a transformational pathway for ensuring sustainable and equal access to the resources and benefit-sharing between humans and nature. The value of nature can be acknowledged

R. A. M. Titumir (✉)
Department of Development Studies, University of Dhaka, Dhaka, Bangladesh
e-mail: rt@du.ac.bd; rtitumir@unnayan.org

R. A. M. Titumir · Md. S. Paran · M. W. Pasha
Unnayan Onneshan, Dhaka, Bangladesh

M. S. Islam
Department of Development Studies, Bangladesh University of Professionals, Dhaka, Bangladesh
e-mail: saeed.islam@bup.edu.bd

R. A. M. Titumir (ed.), *Sundarbans and its Ecosystem Services*, Sustainable Development Goals Series,
https://doi.org/10.1007/978-981-19-3000-3_7

from multiple perspectives which is beyond the singular aspect of monetary valuation, as emphasised by the Intergovernmental Science-Policy Platform on Biodiversity and Ecosystem Services (IPBES). Thus, conceptualising value should take into account the 'plurality of worldviews' of nature since people differ in how they understand the word 'value' while attributing importance to nature (González-Jiménez et al., 2018). Besides, there are different scientific concepts of value—intrinsic, instrumental and relational value that cannot be conceptualised only through market pricing mechanism. As these values are subjected to experience-based and traditional forms of knowledge, exploring human–nature relationships from different angles, along with specific paradigm and methodologies are a crucial way of valuing nature (IPBES, 2018). Thus, IPBES emphasises multiple ways of valuation and how they differ across individuals, contexts and scales (IPBES, 2018). For this, IPBES suggests an accumulated approach with real-world experience and includes local decision-maker (IPBES, 2018). The ultimate purpose is to incorporate the local level knowledge bank with the mainstream policy track, to value natural ecosystems based on multiple grounds.

By breaking down the natural resources into commodities and attaching a price to them, mainstream economics considers the value of nature only in terms of money. Hence, natural resources are presumed under this framework as valuable assets that produce the flow of services to people (Freeman III et al., 2014). This way of valuing nature ignores the intrinsic value of natural resources. Also, in this process, the resources are regarded to be extractable to as much as possible, just for the benefit of human beings that acts as a means of destroying the resources. Moreover, neo-classical economics views environmental pollution and natural resource degradation in terms of 'market failure' and argues that a distorted market cannot guarantee efficient extraction of natural resources. For efficient extraction of natural resources, market factors need to be tweaked to ensure proper distribution of environmental goods and services and these new markets have already appeared. This denotes that the market has the power to regulate the usage of environmental resources efficiently and in a socially optimal way. This is done by initiating curative measures such as levying taxes and providing subsidies that require the intervention of a regulatory body or a government agency (Beder, 2011). Therefore, mainstream economics bats for market-based mechanisms, proposing solutions involving government intervention.

New institutional economics has modified the understanding of the market centric view with new explanations which propose that the over-consumption of natural resources occurs due to a lack of legible property rights. This lack of legible property rights prompts ready availability of cheap resources for end-users which causes the degradation of the resources. The argument is that when property rights are defined, compensations change hands based on the fact as to which party holds the natural resource, ensuring an efficient degree of economic activity. Though this school of thought emphasises the formal institutional arrangement, it does not emphasise enough the informal institutions such as social norms, values and customs which are also crucial for resource conservation and management.

Meanwhile, market-centric approaches fail to understand the dynamic nature of the problem and offer solutions by neglecting the political and economic factors such as power, political settlement and social order. Subsequently, the solutions fail to address these problems. These approaches assume environmental issues as technical problems that can be fixed through market-based mechanisms and technological innovations (Clark & York, 2012). However, in doing so, it ignores the issue of equitable distribution, intergenerational effects and the sustainability of resources. In addition to it, while determining the optimal management of natural resources such approaches neglect humans and their behaviour (Fulton et al., 2011 as cited in Schlüter et al., 2012).

The political and economic approach adopted here helps to deal with the impact of power structures and power relations on usage, management and distribution patterns of natural resources. It argues that an uneven distribution of power causes over-extraction and degradation of natural resources. In fact, the political and economic approach does not explicitly discuss the problem of natural resource degradation. Rather, it offers guidelines to analyse the problem by incorporating some major factors which help to explain as to how the resources are being accumulated for personal gains under a capitalist market economy. Nevertheless, integrating human behaviour into the formal model of natural resource management is still a major challenge (Fulton et al., 2011; Janssen & Jager, 2000; Milner-Gulland, 2012; Schlüter et al., 2012).

Against the backdrop of IPBES-based valuation and critique of different market-centric approaches, this chapter attempts to find how TRUs attach value to the Sundarbans using multiple evidence-based approaches, guided by politics and economy. Taking their perspective into account proves that merely putting a price tag based on the market does not reflect true value of natural resources. Moreover, due to a lack of proper market structure and equal distribution of power, rent through market pricing gets dissipated that goes to the pockets of the rent-seeking powerful class. This rent-seeking behaviour results in unproductive, expropriating activities that generate exorbitant returns for the individuals without any benefit to society. Thus, it has been argued that for conservation and sustainable use of resources, traditional and customary knowledge of the TRUs and their way of valuation should be brought under consideration.

Multiple Values of Nature: Reflections from the Field

According to TRUs, the services received by a human from the forest are of limitless value (Table 7.1). This value cannot be calculated only by market-centric approaches. During FGDs, the TRUs said that high price derived from marketisation of natural resources is causing over-extraction of resources that in turn has been causing serious damage to the forest's ecosystem. Instead of market-based valuation, the TRUs value the forest based on their age-old relationships with the forest.

TRUs' Perspective

All of the respondents argue that the prices of resources are increasing in the market. For which, it has become profitable to collect these resources. Hence, marketing the resources is leading to the intrusion of outsiders, who are not bothered about conservation and sustainability of the resources. Omar Ali, one of the respondents, said:

> Since the market price of resources has increased, the number of resource collectors is also increasing. Now they are coming from outside areas and extracting as much resources as they can. As a result, the forest is losing its resources.

The TRUs live by deploying customary sustainable practices in the forest area over the ages. The way they collect resources and value nature is very different from that of the outsiders and the mainstream valuation perspective (see chapter 4 & 5). Anju Ara, a female respondent, said:

Table 7.1 Services received from the forest

Services	*Example*	*Value type*
Food	Fish, shrimp, prawn, shell, honey, salt, crab, fruits	Direst use value
Water	Transportation	Direct use value
Raw materials	Algae, sand, seaweed	Direct use value
Medicinal resources	Cortex of *Poshur* (*X. mekongensis*), *Hargoza* (*Acanthus ilicifolius*), fruits of *Sundori* (*Heritiera fomes*)	Direct use value
Ornamental resources	Shell	Direct use value
Air purification	Removing toxic elements like SO_2, CO_2 to reduce air pollution	Indirect use value
Protection from disaster	Protection from storm, flood, tsunami etc	Indirect use value
Nutrient cycling	Mineralization of nitrogen and phosphorus by fish through excretion	Indirect use value
Biological control	Ecological balance through maintaining population of wildlife	Indirect use value
Life cycle maintenance	Reproduction of various species of flora and fauna	Indirect use value
Pollination and seed dispersal	Seed dispersal by aquatic animals, water, air and insects	Indirect use value
Habitat	Spawning and nursery grounds for various species of flora, fauna and organisms	Existence value
Hydrological cycle	Movement and storage of water	Indirect use value
Recreation and leisure	Nature watching, sailing, recreational fishing	Direct use value
Aesthetic services	Seascape, landscape, abundance of beautiful biodiversity, etc	Option value
Cultural heritage and identity	Contribution of the Sundarbans to certain cultural tradition, e.g., different songs, dance, rituals, etc. of local community	Bequest value
Future generation possible use	All goods and services which can be used by future generations	Bequest value
Right of existence	Flora, flora and organism that cannot be used but their existence is important to the forest and to the people	Existence value

Source Adapted from Titumir et al., 2019

> Sundarbans not only give us various valuable resources but also it is our intimate relative, our life. We cannot survive a single day without the existence of this forest.

This reiterates the perpetual human–nature relationship and sociality. The respondents believe that valuation for the TRUs is equivalent to life. They consider this mangrove forest as a part of everyday life. However, this straightaway contradicts the mainstream market-centric approach to valuation. Based upon the FGDs, the services and value have been summarised in Table 7.1.

The resources are important for the forest people in many ways. Different parts of flora and different types of fish (and of different sizes of the same kind) are being used by the local people for various purposes (Table 7.2). This proves that forest resource is more valuable for the local people than the commodification entails.

Over-extraction of resources has created a serious threat to the conservation of the biodiversity resources. Therefore, present resource extraction as well as revenue earning is at stake, leading to overvaluation of the current resources by a powerful syndicate.

It is evident that the extraction of most of the forest resources has increased over the years (Table 7.3). As a result, the producer surplus of the forest has reduced. On the other hand, the government is losing revenue due to over-extraction by powerful groups who are grabbing the

Table 7.2 Usage of different parts of a flora

Name of part	*Usage/Importance*
Flower	Honey, fruit, seeds
Fruit	Reproduction of trees, food for birds and animals, food item for human beings such as pickle from *Keora* (*Sonneratia apetala*) fruit
Leaf	Fallen leaves increase the fertility of the forest land, herbal usage, e.g., leaf of *Hargoza*
Root	Prevents soil erosion
Branch	holds leaves to provide shadow
Cortex	Used as herbal medicine, e.g., cortex of *Poshur* works against dysentery
Wood	Making of furniture, boats, pillars of houses, etc
Others	Fish hatch on the roots of the forest trees. Thus, trees help breeding of various species of fish

Source Adapted from Titumir et al., 2019

Table 7.3 Amount of major resources and respective revenue earnings from the Sundarbans during 2001–2002 and 2014–2015

Types of ecosystem services		*2001–2002*		*2014–2015*	
Provisioning	*Produces (unit)*	*Amount*	*Revenue (in US$)*	*Amount*	*Revenue (in US$)*
	Excoecaria Agallocha (*Gewa*) (ft^3)	84,630	33,187	6,026	3,894
	Ceriops Decandra (*Goran*) (no.)	15,865 (MT)	47,742	118,451 (no.)	7,520
	Thatching material *Nypa fruticans* (*Golpata*) (MT)	17,525	33,123	16,868	57,338
	Thatching material grass (MT)	3,621	790	668	225
	Phoenix Paludosa (Hantal)	543 (MT)	348	19,761 (no.)	1,044
	Fuel wood (ft3)	69,370	47,523	14,455	10,190
	Honey (MT)	84	7,970	67	24,048
	Wax (MT)	23	1,665	63	8,108
	Fish (MT)	2,061	58,374	3,432	158,368
	Crab (MT)	123	2,148	1, 123	52,026
	Dry fish (MT)	1,095	18,998	2,773	179,761
Cultural	Tourist (no.)	59,169	14,588	100,817	144,832

Source DoF (2015) cited in Islam and Hossain (2017)

resources and selling them in the market at high prices. Thus, consumers are also facing losses in their accounts. As a result, benefits for the local forest people are decreased.

Analytical Abstraction

When non-marketable goods are turned into marketable goods, it creates rent and deadweight loss. This valuation through the pricing mechanism results in higher prices, loss of consumer surplus, loss in growth and loss in the endowment of the resources (that is, the producer surplus).

Figure 7.1 (in Panel A) explains the same the reality of the natural resource markets. The X-axis indicates price and Y-axis indicates quantity. Initially, at a very low level of price or sometimes with zero price, the demand for natural resources is infinite as measured by the perfectly elastic demand curve. Simultaneously, the earth herself is endowed with a fixed amount of resources. Even though some resources have a regenerative capacity, it depletes when the harvest rate is greater than the regeneration rate. Consequently, the supply curve is perfectly inelastic.

The Panel B of the same figure depicts the effect of price on natural resources if it is transformed into marketable goods. When the non-market goods are transformed into marketable goods, the demand curve will change its position and the consumers will get to interact with

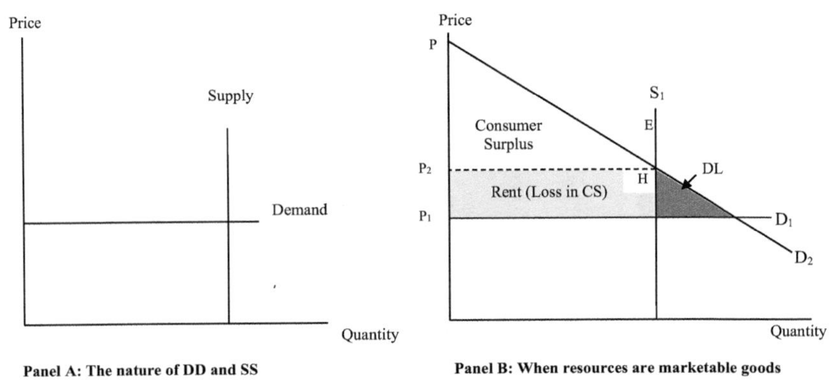

Fig. 7.1 Price effect if the natural resources are turned into marketable goods (*Source* Titumir et al., 2022)

different prices. The changing demand curve produces some alternative distribution and sharing. Firstly, the new equilibrium point sets at E raising the prices from P_1 to P_2. Consequently, the consumer surplus gets reduced from $PEHP_1$ to PEP_2. Thus, loss of consumer surplus as rent is P_2EHP_1. Another important distribution is if the non-marketable goods transform into marketable goods, rent and deadweight loss are generated (shaded area) due to higher demand, and lack of supply causes consumption inefficiency. Consequentially, TRUs are losing their consumer surplus, the government is losing the share of revenue and the forest is losing its endowment of resources. However, the powerful groups are still getting the lion's share of the rent (Titumir et al., 2022).

Summary of Multiple Values Based on IPBES Guidelines

People conceive value based on their everyday interrelations and interdependence with the forest. They often value their resources in terms of what they get from them, both directly and indirectly. The previous classifications of value (Table 7.1) were made based on the way TRUs conceived it. Certain multiple values of nature as conceived by the TRUs are summarised in Table 7.4 as illustrations, using IPBES guidelines.

Rent, Power and Political Settlement

During FGDs with the TRUs, it was revealed that there exists a triangular power distribution mechanism, consisting of people employed by the government agencies (for example, forest department and law enforcers), (illegal) businesses and politicians. They dictate the prices of the resources and control the market by using their political clout. In spite of several rules, regulations and embargoes in place, these powerful groups get extra privilege by bribing the employees of the forest department and police. In some cases, they also get involved in the extraction of resources. However, in the same place TRUs have to get permission to access the forest (for example, boat license, *Golpata* and honey collecting license). These processes require some amount of money. While on the one hand, the forest people have to give an extra toll of money in the form of bribes to the forest department. This forces them to extract more resources from the forest to mitigate their losses. As a result, the biodiversity of resources gets depleted and damage occurs to the ecosystem of the forest.

Table 7.4 An illustrative summary of different meanings of value: collated from the TRUs using IPBES guidelines[1]

	Principles	*Importance*	*Preferences*	*Measures*	*(Explanation)*
Intrinsic	The right of the Royal Bengal Tiger to survive The survival right of the other floral and faunal species	National interest maintaining the world's biggest cat as the world's charismatic mega-fauna and maintaining ecological balance to keep the ecosystem sustainable and resilient	Tigers are natural saviours of the forest and the national symbol of Bangladesh Diverse species maintain the ecological balance that conserves the ecosystem	Tiger Population Census, Narratives, Indigenous knowledge based testimonies, Oral History	Bequest value, Cultural heritage of myths and beliefs, Aesthetic value, World's largest mangrove forest, World heritage site
Instrumental	Ecosystem-based disaster risk reduction and climate regulation, Cyclone protection	Contribution as a natural fortress in reducing natural disaster and contribution of carbon sequestration to the global stock	Interest in maintaining habitat and biodiversity of this diverse ecosystem	Quantitative, Narrative, Indigenous knowledge-based testimonies, Oral History	Resist soil erosion, combat salinity intrusion and provide fresh air
Relational	Living in harmony with nature, Customary Rights, Rights to self-determination, Sustainability and resilience	Identifying TRUs as key to conservation and sustainable use of biodiversity for sustainability and resilience	Interest in maintaining these multiple ecosystems that provide different ecosystem services (provisioning, regulating and supporting)	Economic price-based measures are inadequate and following are to be explored: Narrative, Indigenous, Deliberations, Multiple evidence-based approaches	Provides livelihood services (food, water, fuel, etc.), Identity and autonomy, Living well in harmony with nature, Spirituality of nature, cohesion, Governance and justice

Source Adapted from Titumir et al., 2019

[1] These are examples and itself could be an article if an attempt is made to capture it in its entirety.

On the other hand, if powerful groups get caught by the forest people violating any rule, they immediately get released by paying the bribe. In this context, local forest people have mentioned such groups as the *black party* that extract resources regularly, violating the laws. If anybody somehow catches them violating the law, they seize that person and do not release them till they are done with illegal extraction. Thus, rent dissipation and uneven power distribution have caused a clientelist network through which extraction of resources occurs randomly to accumulate more rent. This acts as a major cause of the degradation of the ecological biodiversity of the forest (see Titumir et al., 2022).

Governance Structure and Sign of Unsustainable Resource Extraction

The Sundarbans was announced as a reserve forest in 1875. About 32,400 hectares of this forest have been declared as three wildlife sanctuaries. These wildlife sanctuaries were established in 1997 under the Bangladesh Wildlife Preservation Amendment Act, 1974. Afterwards, it was declared as an Ecologically Critical Area (ECA) by the Department of Environment (DoE) of Bangladesh in 1999, under the Bangladesh Environment Conservation Act, 1995, which was amended in 2010. According to the conservation act 2010, ECA refers to an area that is rich in unique biodiversity or due to its importance from the environmental perspective, becomes necessary to protect or conserve from destructive activities. ECA also falls within the category of natural and cultural heritage.

The protected area of the Sundarbans helps maintain the ecological harmony and continue the sustainability of biodiversity and ecosystem management. The Sundarbans reserve forest not only contributes to enrich its floral and faunal biodiversity in both terrestrial and marine ecosystems, but also gives tremendous protection to coastal populations from cyclones, tidal floods, storms, erosion and other natural calamities (Das & Vincent, 2009; Payo et al., 2016 cited in Islam & Bhuiyan, 2018). The ECA works as a protective barrier or natural wall for the coastal communities of the Sundarbans (Islam, 2013; Bhowmik & Cabral, 2013 cited in Islam & Bhuiyan, 2018).

Despite being declared as ECA, the rules as per guidelines required to protect the ECA were not implemented and most importantly the rights of the original inhabitants of the forest are equivocal and trifling. As a result, opportunist groups took advantage and are still causing over-exploitation of the forest resources. Problems have appeared due to lack of participation of the TRUs in the conservation process, declination of user rights, lack of implementation of laws and rules, the emergence of imposed development projects and use of the top-down approach in the co-management of forest resources. As a result, decisions are mostly made at the authority level and are transmitted to the lower level without substantial discussion and considering its impact. Meanwhile, due to a lack of recognition for traditional knowledge and other kindred circumstances, the significance of being an ECA has been smothered to a great extent. As mentioned earlier in this book, apart from *Rampal*, recently another electricity power plant has been planned to be established only 10 km away from the ECA area. According to the environment law of Bangladesh, establishing a power plant within 10 km of an ECA zone is prohibited.

Apart from the problematic rules and implementation scenario with regard to ECA management, it is also to be noted that the specified sanctuary areas cover less than 1/3rd of the total area of the Sundarbans in Bangladesh and in case of sustainable use/overall protection, the ECA based strategy is not sufficient. On this backdrop, the absence of holistic and participatory conservation strategies coupled with several socio-economic pressures has put the natural ecosystems of the Sundarbans at jeopardy.

Despite all these objections in place, rapid industrialisation is happening in violation of the existing laws on the periphery of the forest. Hence, it is clear that the legal remedies and existing governance structure have proven to be ineffective when it comes to the implementation of laws. As a result, the conservation strategy is being hampered while sustainability of the forest is at stake due to the relentless exploitation by the powerful groups and lackadaisical attitude of the administration.

TRUs' Perspectives

The respondents said that their resource collection processes don't intervene with the conservation process of the forest. Abu Musa, one of the respondents said,

> TRUs usually collect resources keeping in mind that if they cause any harm to the forest, they will be the ones who would suffer.

However, due to faulty execution of these laws, rules and regulations, access to the forest had become restricted for the TRUs but the powerful people kept on extracting resources unhindered by bribing the police and forest department officials. Abu Musa further added,

> As the powerful class is getting access to the forest bribing a huge amount of money, they are extracting as many resources as they can without caring about the survival of the forest. But TRUs are getting deprived.

Even though clientelist groups engage in resource extraction employing unsustainable practices, their illegal activities go unpunished while the TRUs face the brunt and are harassed by the authorities as they are powerless and have no participation in the decision-making process of managing the forest resources. Tanjila, a female TRU says,

> They (forest department and other governance bodies) do not listen to us. They only pay heed to the opinion of the powerful class. We, the TRUs, are the victims.

During the FGDs it was known that the rent-seeking groups who are politically and bureaucratically powerful, use illegal means to extract resources. Moreover, a group of people was even found to be using poison and prohibited nets to catch more fish at a single go which in turn destroys all kinds of fish along with other aquatic species (Fig. 7.2).

Figure 7.2 shows the negative impact of poisoning and using prohibited nets for catching fish, which leads to the loss of biodiversity, damages spawning of eggs, leads to losses in both consumer and producer surplus and more importantly also destroys fish and other aquatic fauna. Despite all this, the process is carried out only for benefiting the powerful people.

A few studies show that the existence of organised groups of mongers, who illegally cut and remove valuable trees, is prevalent in the forest region (Rahman et al., 2010). These groups collect fish using illegal means by bribing the forest department people (Hassan et al., 2012). Moreover, the existence of vertical relations in society and upward enforcement of rules enable the powerful group to capture resources with liberty (Adhikari & Goldey, 2010). Due to unequal distribution of power and wealth, conflict and discrimination among different stakeholders arise (Hassan et al., 2012). Therefore, biodiversity degradation not only occurs due to the non-existence of markets but also because of unequal power

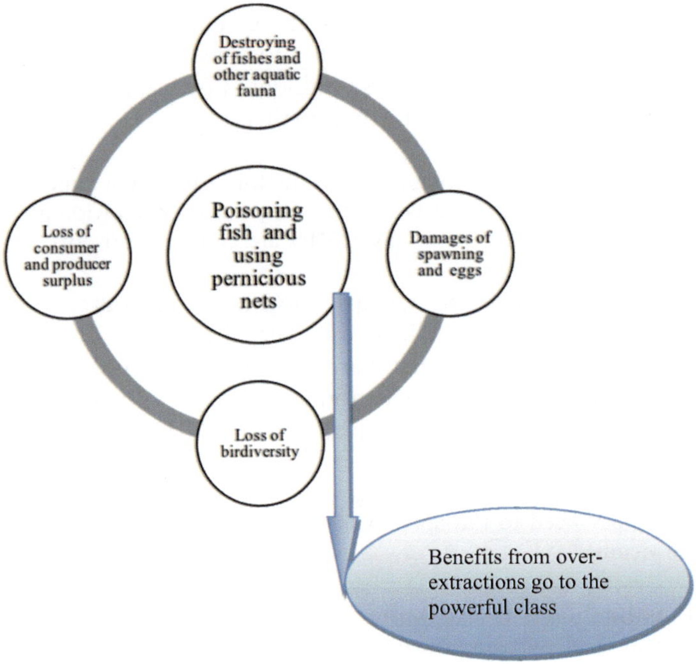

Fig. 7.2 Effects of poisoning and using prohibited net for catching fish (*Source* Adapted from Titumir et al., 2019)

distribution among different groups. In addition, infiltration and illegal removal of valuable wood from the forest occurs due to the absence of sustainable management practices and properly functional institutional arrangements (Rahman et al., 2010).

Analytical Abstraction

The agents claim control over the potential rent under the extractive institutional arrangements that are historically prevalent. This is not solely dominated by the objective of production but also by the prevailing social relations and the scale of production, along with the relations of distribution and property regimes. Specifically, it can be said that in a developing economy, the likelihood of unstable property rights over natural resources is very high. A strong institutional arrangement can check the stability.

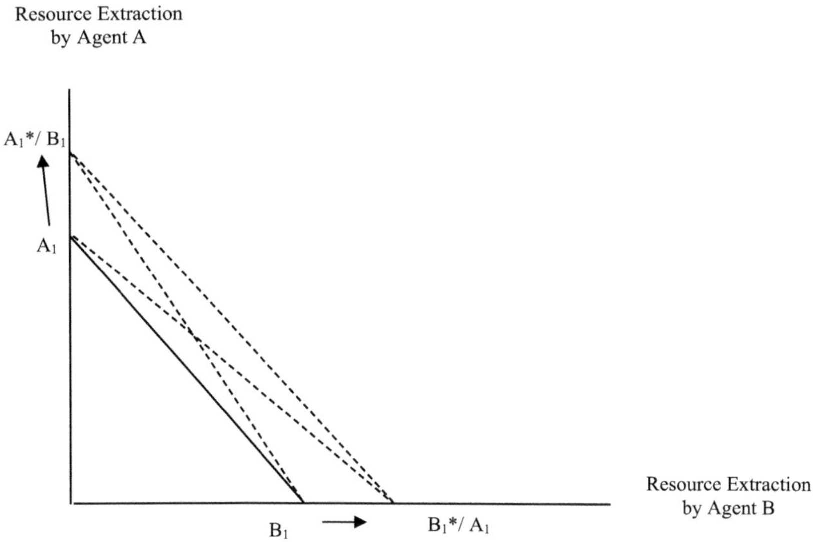

Fig. 7.3 Institutional vulnerability and destruction of resources (*Source* Titumir et al., 2022)

Here, it has tried to visualise how different agents behave strategically under the vulnerable institutional arrangements. This is to serve their own interest under the game-theoretic approach which ultimately results in the destruction of natural resources.

The phenomenon of the graph in Fig. 7.3 can be very useful to depict the prisoner's dilemma and the Nash equilibrium[2]; in addition to their rational incentive to break the contract under the weak institutional arrangement. Here, the vertical axis measures the resource extraction by agent A and the horizontal axis measures the resource extraction by agent B. Under the mutual contract, both the agents extract resources on the A_1B_1 line. The next scenario is that as a rational being, agent B can maximise if B thinks that holding agent A constant, he (B) can extract more resources as there is no one to monitor. Hence, agent B will go

[2] Nash equilibrium is a steady-state where each agent has no incentive to deviate from their chosen strategy assuming the strategy taken by the opponent.

for more extraction at B_1* keeping the A_1 fixed which is a symptom of cheating, pointing B_1*/A_1 in the graph. Similarly, another rational agent of this model A will do the same at A_1* keeping the B_1 fixed is another symptom of cheating, pointing A_1*/B_1. Therefore, under a weak institutional arrangement, the contract does not hold. In the graph, the new resource extraction line is now A_1*/B_1 and B_1*/A_1 galloping up from A_1B_1. Despite being Nash equilibrium, it is unstable. Until the complete extraction of natural resources occurs, the shift and alteration of the non-cooperative game will remain. The key underlying essence of this proposition is that the faulty persistence and the adverse development, of the institutions of the ex-colonial country, is the reason for massive resource destruction (Titumir et al., 2022).

Towards Transformational Pathways

Transformational pathways require stabilisation of the ecosystem that can be achieved by conservation of the forest resources by lowering the damage done to it. For stabilising the ecosystem, it is necessary to identify the drivers of ecological degradation. These drivers include both natural and anthropogenic pressures on the forest. Moreover, institutional and governance structures are involved in the degradation process by allowing various means of exploitation of natural resources, despite being the supposed saviour of the forest. Since indigenous local knowledge has been kept outside of this governance and institutional framework, it is crucial to incorporate it into the stabilisation process. This stabilisation will push biodiversity and the ecosystem in a situation, where sustainable use of resources and benefit-sharing among humans and nature will be ensured. Conservation and restoration of nature will lead to an ambience of well-being for both humans and nature (Fig. 7.4). When both live in harmony, the ecosystem becomes balanced through sustainable use, access and benefit-sharing between them.

If the transformation of an ecosystem cannot be achieved, there will be an imbalance, leading to the destruction of the biodiversity, which is elaborated in the following three scenarios.

Scenario A: Loss of Biodiversity (Level of Alienation)

The TRUs questioned the misconception of humans that prompts them to consider themselves as independent and master of nature, despite the

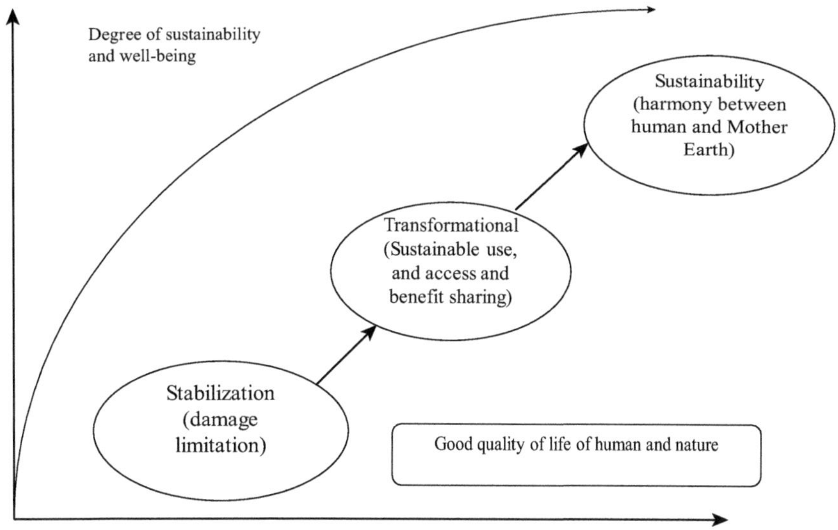

Fig. 7.4 Transformation pathways towards sustainability (*Source* Based on Titumir et al., 2019)

fact that they themselves are a part of nature (see Titumir & Paran, 2022a, 2022b, 2022c). They said that generally, people treat nature as 'mere matter' that can be extracted for human use, usually by destroying the natural resources in various ways. In this way, according to them, human beings separate themselves from nature. People become separated from the world when they fail to recognise its humanity and are unable to see the world as their own and themselves as part of the world. Thus, this break-up derives from people's failure to understand the sociality between humans and nature. These broken relations, as they suggest, lead to the commodification of natural resources through market pricing. In this way, valuation becomes equal to the market price and the intrinsic value of natural resources and the sociality between nature and humans gets neglected. This gives rise to over-extraction resulting in the destruction of natural resources (Fig. 7.5). On the contrary to the virtuous cycle of the ecosystem, the vicious cycle is drawn by producers, consumers and decomposers, as all of them start to get away from each other and start to bear losses in each of the provisions (see Titumir & Paran, 2022a).

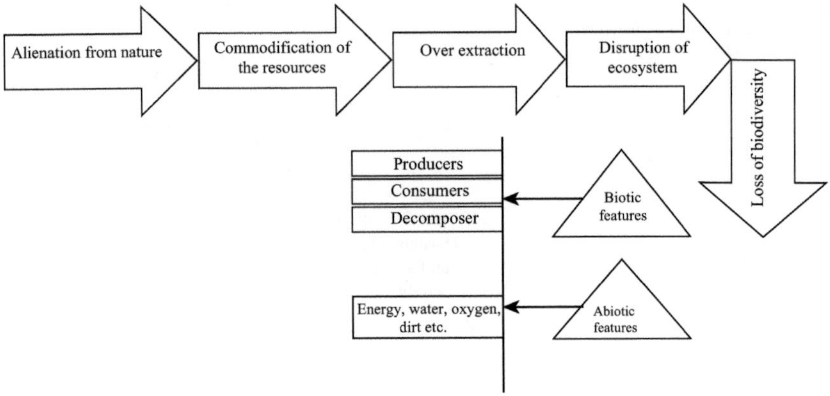

Fig. 7.5 Alienation and loss of biodiversity due to over extraction and disruption of ecosystem (*Source* Based on Titumir et al., 2019)

Scenario B: Loss of Biodiversity (Commodity Fetishism)

The TRUs keep emphasising on pervasive commodification of natural resources. Everything produced or received from nature is seen as a commodity in their mind. If this prevails, commodities will soon turn into fetishes in the sense that human belief in commodity will create an obscure hierarchy of value of the natural resources, upon which the demand of commodity will depend. The high price commodities are considered to be the most valuable ones. This commodity fetishism induces over-extraction of the resources through the primitive accumulation of nature (Fig. 7.6). The greater the money made by the powerful class; the greater will be the influence it can put on extracting resources. Through this pervasive motivation, according to the TRUs, a clientelist network emerges wherein members of the network thrive through primitive accumulation. Through this, only material provision of the forest comes forward to condone the intrinsic and underlying values of mutual existence (Titumir et al. 2022a). Degradation of the forest, loss of biodiversity and overall socio-ecological imbalances occur.

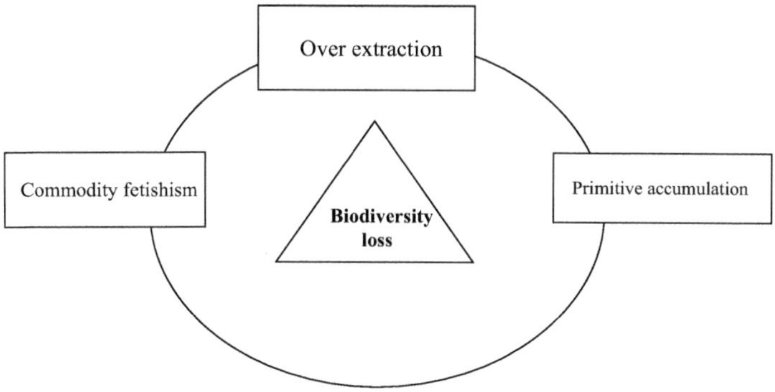

Fig. 7.6 Commodity fetishism and biodiversity loss (*Source* Based on Titumir et al., 2019)

Scenario C: Sustainable and Non-sustainable Scenario (Appropriation and Expropriation)

As stated earlier, this separation occurs through the materialistic provision of commodities where the capitalist class tends to get access to the resources and sell them in the market at a higher price. By using the power of the institution, powerful agents confiscate the resources. The extractive institutions exclude the TRUs and the primary producers to equitable benefit-sharing of natural resource and the state also loses a significant share of revenue. The TRUs claim that the officials of the local forest departments, police and leaders of political parties 'successfully' marginalise the local people or community.

In contrast to this non-sustainable use of the resources, proper appropriation of the natural resources of the forest is healthy. In addition to this, it is important to strengthen the customary and traditional rules and practices of the forest people, who genuinely treat the forest as their mind. The key understanding is that these socially constructed norms and values which are the informal institutional set up can solve the natural resource problem. The norms and values formulate the collective organisation that can preserve the natural resources in a sustainable and equitable manner (Titumir & Afrin, 2017; Titumir et al., 2020). The authority to impose credible threats and the sanction by the stakeholders of the resources on the resource distribution can excessively solve the natural resource problem and ensure the distribution of benefits. As forest people regard the forest as their intimate relative, they do not think of harming the natural resources of the forest. Therefore, their values, norms and traditional rules and practices should be incorporated into the conservation process.

Conclusions

This chapter took into account the multiple values of nature along with the transformational pathways. These values refer to different types such as intrinsic, instrumental and relational. These values were also emphasised by IPBES, along with some suggested and accumulated approaches based on real-world experiences. Moreover, it has also been revealed that these values, especially the true value of the forest cannot be proven or reflected through mere market pricing. This market-centric approach cannot determine its true worth. This is because the value of the forest lies truly in their age-old relationship, from the perspective of the TRUs. Respondents also believe that this valuation for the TRUs is equivalent to life. TRUs consider this forest to be their mind through which human–nature sociality flourishes.

They challenged the mainstream valuation methods and argued that the valuation of environmental resources based on market pricing, does not look into the social benefits and values coproduced through complementarity between human and nature. Their reasoned knowledge and practices, documented and demonstrated, cast a shadow upon the conventional view of sustainable governance. This is done as an abstract trade-off between human activities or environmental protection which is based on a Cost–Benefit Analysis (CBA). This presumes that humans are the external agents to the governance of natural resources. In this process, any monetary tag fails to reflect the social opportunity costs that may arise from the dependence of the community on the resources for income, jobs and livelihoods. Overall, the value of the production network within the socio-ecological production landscape remains undervalued in the market-based valuation system.

The TRUs argue that the weak market structure and unequal distribution of power cause rents to be dissipated which goes into the pockets of the rent-seeking powerful class. As a result, rent-seeking agents comprising bureaucrats, businessmen and politicians extract resources in an exploitative manner which destroys biodiversity. These unproductive, expropriating activities might bring positive returns to the individual but the same is not for society. Finally, the TRUs indicate that the traditional knowledge system can significantly contribute to the sustainable management of biodiversity of resources. This can occur both within the protected area system and potentially as other effective area-based conservation measures considering that they are given an opportunity and are supported by the government and non-government agencies.

Moreover, some analytical abstraction helped to reveal that when non-marketable goods are turned into marketable goods, it creates rent and deadweight loss. It also tends to lower the consumer surplus of the TRUs, causes the government to lose out on shares on revenue and leads to improper use of resources.

Acknowledgements The chapter draws on 'The Sundarbans is our mind: an exploration into multiple values of nature in conversation with traditional resource users', published by United Nations University Institute for the Advanced Study of Sustainability, Tokyo, Japan, 2019, in *Understanding the multiple values associated with sustainable use in socio-ecological production landscapes and seascapes (SEPLS), Satoyama Initiative Thematic Review, vol. 5.*

References

Adhikari, K. P., & Goldey, P. (2010). Social capital and its "downside": The impact on sustainability of induced community-based organizations in Nepal. *World Development, 38*(2), 184–194.

Beder, S. (2011). Environmental economics and ecological economics: The contribution of inter disciplinarity to understanding, influence and effectiveness. *Environmental Conservation, 38*(2), 140–150.

Clark, B., & York, R. (2012). Techno-fix: Ecological rifts and capital shifts. In A. Hornborg, B. Clark, & K. Hermele (Eds.), *Ecology and power: Struggles over land and material resources in the past, present, and future*. Routledge.

Freeman, A. M., III., Herriges, J. A., & Kling, C. L. (2014). *The Measurement of environmental and resource values: Theory and methods*. RFF Press.

Fulton, E. A., Smith, A. D. M., Smith, D. C., & van Putten, I. E. (2011). Human behavior: The key source of uncertainty in fisheries management. *Fish and Fisheries, 12*(1), 2–17.

González-Jiménez, D., Berghöfer, U., Berghöfer, A., Heubach, K., Kosmus, M., & Bertrab-Tamm, A. V. (2018). *Beyond measurements: Multiple values of nature and their diverse conceptualization*. Intergovernmental Science-Policy Platform on Biodiversity and Ecosystem Services (IPBES).

Hassan, S. M. M., Un-Nabi, M. R., & Mozumder, M. (2012). Resources utilization of Sundarbans mangrove forest people: An ecological perspective. *Asian Profile, 40*(4), 297–309.

Islam, M., & Hossain, Mm. (2017). Community dependency on the ecosystem services from the Sundarbans mangrove wetland in Bangladesh. In A. K. Prusty, R. Chandra, & P. A. Azeez (Eds.), *Wetland science: Perspectives from South Asia*. Springer.

Islam, S. M. D., & Bhuiyan, M. A. H. (2018). Sundarbans mangrove forest of Bangladesh: Causes of degradation and sustainable management options. *Society for Environmental Sustainability, 1*, 113–131.

IPBES. (2018). *Beyond measurements: Multiple values of nature and their diverse conceptualization: An introduction for practitioners, based on the IPBES preliminary guide on multiple values.* https://web2.questback.com/media/fd.ashx?mid=799094&guid=%7B6153EB85-0660-430C-950D-703B374A4126%7D. Last accessed on 13 March 2021.

Janssen, M. A., & Jager, W. (2000). The human actor in ecological-economic models. *Ecological Economics, 35*(3), 307–310.

Milner-Gulland, E. J. (2012). Interactions between human behaviour and ecological systems. *Philosophical Transactions of the Royal Society of London, Series B, 367*, 270–278.

Rahman, M. M., Rahman, M. M., & Islam, K. S. (2010). The Causes of deterioration of Sundarbans mangrove forest ecosystem of Bangladesh: Conservation and sustainable management issues. *Aquaculture, Aquarium, Conservation & Legislation International Journal of the Bioflux Society, 3*(2), 77–90.

Schlüter, M., Mcallister, R. R. J., Arlinghaus, R., Bunnefeld, N., Eisenack, K., Hoelker, F., Milner-Gulland, E. J., & Mueller, B. (2012). New horizons for managing the environment: A review of coupled social-ecological systems modeling. *Natural Resource Modeling, 25*, 219–272.

Titumir, R. A. M., & Afrin, T. (2017). Complementarities of human-nature well-beings: A case illustrated through traditional forest resource users of Sundarbans in Bangladesh. In *Satoyama Initiative Thematic Review (SITR)—Volume 3*. United Nations University Institute for the Advanced Study of Sustainability (UNU-IAS).

Titumir, R. A. M., Afrin, T., & Islam, M. S. (2020). Traditional knowledge, institutions and human sociality in sustainable use and conservation of biodiversity of the Sundarbans of Bangladesh. In O. Saito, S. M. Subramanian, H. Hashimoto, & K. Takeuchi (Eds.), *Managing socio-ecological production landscapes and seascapes for sustainable communities in Asia: Mapping and navigating stakholders, policy and action* (pp. 67–92). Science for Sustainable Societies. https://doi.org/10.1007/978-981-15-1133-2_5

Titumir, R. A. M., Afrin, T., & Islam, M. S. (2022). *Natural resource degradation and human-nature wellbeing: Cases of Biodiversity resources, water resources, and climate change*. Springer (Forthcoming).

Titumir, R. A. M., & Paran, M. S. (2022a). Human-Nature Cooperation for Well-Being: Community Understanding on One Health Approach in the COVID-19 Era in the Sundarbans. In M. Nishi, S. M. Subramanian, H. Gupta (Eds.), *Biodiversity-Health-Sustainability Nexus in Socio-Ecological Production Landscapes and Seascapes (SEPLS)*. Springer Nature, Singapore.

Titumir, R. A. M., & Paran, M. S. (2022b). Ecosystem services and well-being in the Sundarbans of Bangladesh: A multiple evidence base trajectory. In R. Dasgupta, S. Hashimoto & O. Saito(Eds.), *Assessing, mapping and modelling of mangrove ecosystem services*. Springer Sustainability Science Book Series.

Titumir, R. A. M., & Paran, M. S. (2022c). Human-nature relationship for post 2020 biodiversity framework: An exploration into traditional knowledge, customary sustainable use and community-based innovation in the Sundarbans. In R. A. M. Titumir (Ed.), *Fifty years of nation building: Political economy of Bangladesh's development*. Centennial Book Series Project of University of Dhaka. Department of Development Studies, University of Dhaka.

Titumir, R. A. M., Paran, M. S., & Pasha, M. W. (2019). The Sundarbans is our mind: An exploration into multiple values of nature in conversation with traditional resource users. In UNU-IAS & IGES (Eds.), *Understanding the multiple values associated with sustainable use in socio-ecological production landscapes and seascapes (SEPLS)* (Satoyama Initiative Thematic Review, Vol. 5, pp. 97–117). United Nations University Institute for the Advanced Study of Sustainability.

CHAPTER 8

Climate Change and Its Impact: Sundarbans as a Natural Wall

Rashed Al Mahmud Titumir, Md. Shah Paran, Mostafa Walid Pasha, and Mahtarin Haq Meem

Introduction

This chapter explores the impact of climate change in the Sundarbans and studies it from the context of how the largest composite mangrove forest in the world is responding to climate change. It brings forward how the forest is an ideal example of an ecosystem that works as a natural wall against climate change. Accordingly, it depicts the significant role the forest plays in tackling climate-related challenges for Bangladesh.

R. A. M. Titumir (✉)
Department of Development Studies, University of Dhaka, Dhaka, Bangladesh
e-mail: rt@du.ac.bd; rtitumir@unnayan.org

R. A. M. Titumir · Md. S. Paran · M. W. Pasha · M. H. Meem
Unnayan Onneshan, Dhaka, Bangladesh

R. A. M. Titumir (ed.), *Sundarbans and its Ecosystem Services*, Sustainable Development Goals Series,
https://doi.org/10.1007/978-981-19-3000-3_8

Recently, climate change has become a burning issue as the natural green cover is deteriorating steadily given several anthropogenic pressures caused by humans. In response to this, the Sustainable Development Goals (SDGs) were drawn which also highlights 'climate action' in Goal 13. Numerous anthropogenic drivers are responsible for the degradation of the natural ecosystem, leading to climate change. Rapid urbanisation gives way to mindless industrialisation constantly disregarding the necessity and susceptibility of nature. Carbon emission is skyrocketing by the day scoring record increases every year; as a result, the ozone layer is facing the brunt of being depleted at a fast pace. To add to this, air, water and soil pollution are also on the rise. The benefits of industrial development marketed as growth have overthrown the desire for environmental sustainability, as a result of which, global warming has gone up melting massive icebergs and glaciers, with incidents of avalanches increasing along the sea. Several species of flora and fauna are becoming extinct. Nature, however, has been dealing with co-morbidity for years.

Globally, natural disasters arising from climate change are on the rise resulting in a loss of around $142 billion every year since the last decade. Bangladesh has been categorised as one of the most vulnerable countries impacted by climate change. Almost 80% of Bangladesh is comprised of the deltaic floodplain of the Ganges–Brahmaputra-Meghna (GBM) basin. It is, however, more vulnerable to tropical cyclones and is ranked sixth for being the most vulnerable to floods (Government of Bangladesh, 2008). Millions of lives and livelihoods are affected by frequent natural disasters caused due to climate change. The changes in climate variables will be devastating in the upcoming years (Table 8.1). Bangladesh, however, is

Table 8.1 Climate change scenario in Bangladesh

Year	*Temperature change °C mean*		*Precipitation change % Mean*		*Sea level rise (cm)*		
	Monsoon season	*Dry monsoon*	*Monsoon season*	*Dry monsoon*	*SMC*	*NAPA scenario*	*IPCC 3rd Upper range*
2030	0.8	1.1	+6.0	−2.0	18	14	14
2050	1.1	1.6	+8.0	−5.0	30	32	32
2100	1.9	2.7	+12.0	−10	60	88	88

Source Climate Change Cell (2008)

fortunate to have the forest located in its south-west coastal region which acts as a natural shield protecting it from such natural disasters.

The impact of climate change on nature, life and livelihoods in various forms is devastating. The rising salinity of water in the coastal and marine ecosystems is negatively affecting the growth of mangroves, leading to severe crop failure and reduced crop yields. The salinity intervention in agricultural fields is mainly due to high tides, cyclones and storm surges, prolonged saltwater stagnation and flood (Chowdhury & Brahma, 2019) which are impacts of climate changes. The saltwater intrusion into the surrounding agricultural lands has resulted from shrimp culture (Flaherty & Vandergeest, 1998; Flaherty et al., 2000) and a rise in the sea level.

As a consequence of increased salinity, the major mangrove species are dying. About 5–6% of '*Sundari*' (*Heritiera fomes*) trees are affected by 'Top Dying' disease caused by increased salinity. Almost 50% of the '*Sundari*' trees in high salinity zones are affected by this disease (Rahman et al., 2010).

This threatening journey through uncertainty and natural catastrophe cannot be resisted if humans fail to recognise the perpetual importance of the natural ambience and its elements. This chapter explores the role played by the mangroves in fighting against storms and cyclones, carbon emission, salinity intrusion, tidal surge, sea-level rise and other such natural hazards.

Impact of Climate Change: Perception of TRUs

To understand the perception of traditional resource users in the region, regarding the impacts of climate change on nature, alongside their life and livelihoods, this study uses the Likert Scale.[1] Total 14 statements on a five-point scale were provided as a survey questionnaire that directly or indirectly represents the impact of climate change. The five-point scale asked the respondents if they strongly agree, agree, have no opinion, disagree or strongly disagree with each of the statements provided. The number of respondents was 115 and they were chosen randomly. The

[1] A survey scale was developed by Rensis Likert in 1932 for measuring attitude. It is widely used to measure the strength/intensity of attitude, perception, opinions, that is, degree of perception ranging from strongly agree to strongly disagree with a statement representing an issue/event/case.

respondents are the members of the *Koyra Multipurpose Cooperative* who pursue their livelihoods as TRUs. They are regular visitors to the forest and depend on it for survival.

Most of the respondents argued that the salinity in the coastal areas is increasing, affecting both life and livelihoods. Around 70% of the respondents agreed that salinity in the rivers has increased, while 18% of them strongly agreed with this statement. Around 10% of the respondents, strongly disagreed implying that water in the rivers was always saline and they do not see any change (Table 8.2).

Along with increasing salinity, water in the rivers is getting blackish gradually. Total 73% of the respondents agreed that the river water is turning black and dark. Similarly, water in ponds and tube-wells is also becoming saline. Around 72% of the respondents agreed that they are dependent on rainwater for potable use. Around 83% of the respondents agreed that salinity in soil has also increased. As a result, crops and trees are dying.

The impact of salinity intrusion in the region can be depicted by watching the pattern of decreasing crop yield in agricultural lands. Around 65% of the respondents disagreed with the statement that the yield of crops has increased. Nonetheless, it is worthwhile to mention that 30% of the respondents strongly agreed and 62% agreed that shrimp culture has increased in the region which is regarded as one of the main factors contributing to the rise in salinity of water and soil (Table 8.2).

During the survey, it was found that the availability of fish in the rivers has decreased in number (see chapter 3 also). However, the respondents argued that they are still catching a variety of fish from the rivers. The extraction of resources (honey, fish and crabs) from the forest has increased. The decline in the availability of fish despite many varieties found in the region is the result of over-extraction. The distribution of trees and *golpata*, unfortunately, has not increased. Around 78% of the respondents disagreed and 20% of the respondent strongly disagreed with the statement that the number of trees and *golpata* in the forest has increased (Table 8.2).

Land Use and Land Cover (LULC) Predication

This study adopted the Maxent model (v.3.4.1) to see future geographic distributions in the year 2050 of various floral species (big trees) in the Sundarbans's Koyra region. This model is built on deterministic

Table 8.2 Impacts of climate change: TRUs' perception

Impact	*Total*	*Strongly agree (%)*	*Agree (%)*	*No opinion (%)*	*Disagree (%)*	*Strongly disagree (%)*
Salinity increased in river water	115	18	70	1	0	10
Blackish water in the rivers	115	1	73	2	23	1
Saline water in the ponds	115	1	62	5	30	2
Saline water in tubewells	115	10	65	7	10	7
Soil became saline	115	1	83	6	10	0
Dependent on conservation and drinking of rain water	115	16	72	6	5	1
Crops yields increased	115	0	25	7	65	3
Fish increased in rivers	115	12	17	5	64	1
Variety of fish caught in the rivers	115	12	57	6	24	0
Shrimp culture increased	115	30	62	9	0	0
Amount of honey collection increased	115	11	73	4	10	2
Amount of fish collection increased	115	8	26	4	61	1
Amount of crab collection increased	115	11	48	3	36	2
Number of trees and golpata increased	115	0	0	2	78	20

Source Authors' calculation

algorithms covering optimal (maximum entropy) probability distribution (Phillips, 2017; Phillips et al., 2006). It adopts only current data of species distribution to predict the links of species coverage and environmental factors (Elith et al., 2011; Marcer et al., 2013).

ArcMap (v.10.2; ESRI, Redlands, CA, USA) was used to integrate the GPS and remote-sensing data. The number of variables for environment was 12 which include climatic and physical condition, soil conditions and anthropogenic factors. Precipitation and air temperature have been identified as climatic data. Geomorphological data include elevation with a spatial resolution of 9.99 m. Data on anthropogenic pressures have been collected by measuring distance of road and human settlement with 500 m altitude from the ground, and Google Earth Pro software was used to digitise and save them as Kml files. The Kml files were changed to raster grids using ArcMap software (v.10.2; ESRI). Soil nutrient variables include nitrogen (N), phosphorus (P), Organic carbon (OC), potassium (K), calcium (Ca), pH and organic matter (OM). Based on the same geographic boundary of World Geodetic System 1984 longitude-latitude projection (resolution: 9.99 × 9.99 m) all of these data were altered into ASCII format in raster grids. The bio-climatic factors include mean annual temperature, annual precipitation, annual range in temperature and precipitation and the temperature of the coldest and warmest month as well as precipitation of the wet and dry season.

Contribution of the Factors

The predictions suggest that current trends of climatic variables, geomorphology, soil nutrients and anthropogenic factors will lead to massive degradation of the forest coverage (almost half) in 2050. As a result, almost half of the forest in this region will become empty fallow land by 2050 (Fig. 8.1). Three bio-climatic variables with the highest contribution are mean air temperature in the coldest season, precipitation in the driest month and mean air temperature in the wettest season. The forest in this region are at higher risk to be impacted by anthropogenic pressures.

Conceptualising the Mangroves as a Natural Wall to the Climate Effect

Predominantly, climate change includes: the rise in temperature, changes in weather patterns and rise in sea level. These changes have direct and indirect impacts on the ecosystem and humans. Climate change causes distortion in the hydrological system, salinity intrusion, warmer summers and irregular rainfall, increased recurrence of natural disasters, displacement of people, biodiversity loss, poverty and health hazards, loss of food security and agricultural production as well as conflict and violence. Well-established forests such as the mangroves and tree plantations offer

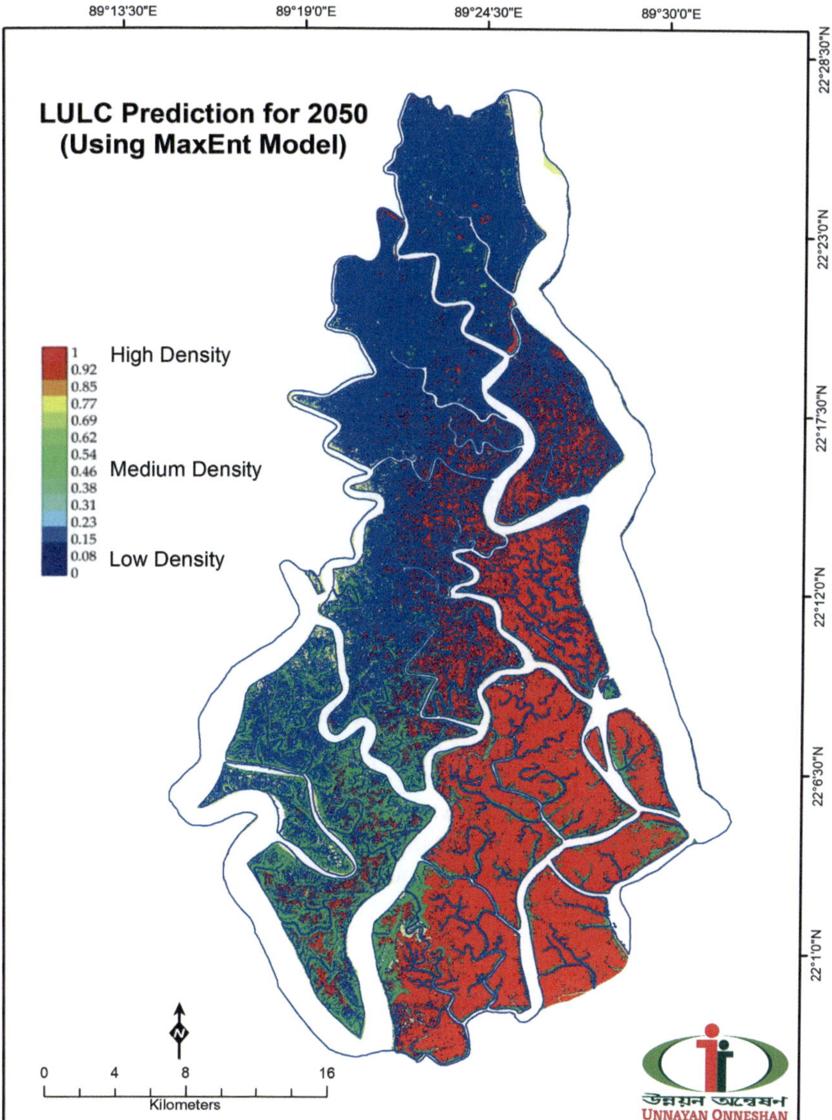

Fig. 8.1 LULC prediction using MaxEnt model (*Source* Unnayan Onneshan [2020])

an effective physical barrier against the impact of climate change saving both lives and property. Roots of the mangroves trap sediments and help build coastline by stalling erosion. Mangroves works against tidal surges, wave and storms like other wetlands as horizontal barriers. This type of ecosystem can curtail the height of waves and control the catastrophic impact of tsunami. The natural barriers to disasters are more effective and cost-saving compared to man-made systems and saves money, lives and livelihoods.

One of the important ecosystem services rendered by mangroves is the ability to alleviate damages of natural disasters (Barbier, 2016; Das & Vincent, 2009; Marois & Mitsch, 2015). Mangroves protect coastal areas from waves and currents (Granek & Ruttenberg, 2008), playing an important role in protecting people on the coast from tropical cyclones (Spalding et al., 2014). The protection of the mangroves is very important since people and communities in coastal areas are getting more vulnerable to increasing climate shocks on a daily basis (Marois & Mitsch, 2015; Woodruff et al., 2013). However, these services from the mangrove ecosystem are generally undervalued regardless of its great contribution to the coastal communities (Badola & Hussain, 2005; Brander et al., 2012).

The existence of mangroves in coastal settings provides them with the ability to cope with natural hazards. This is because mangroves exist in those areas by recovering from the negative impacts of such hazards. Apart from this, it also provides multiple benefits and services that are crucial to moving beyond the vulnerabilities due to climate change. It is found that surface of soil in mangroves is affected by both physical and biological process, such as water flow, sedimentation, plant debris and roots etc. Aerial roots of mangroves such as prop roots, pneumatophores, knee roots and plank roots keep the soil intact and reduce erosion (Krauss et al., 2014). Mangroves helps land maturation and works against vulnerability to cyclones and natural hazards by forming sedimentation, minimising exposure to waves and developing peat. For example, sediments at the mouth of the Amazon river get trapped in the mangroves of French Guiana (FAO, 2007).

Causing harm to the mangroves has an adverse impact on the ecological cycle of the forest's biodiversity. This further affects human beings and other floral and faunal species fatally. Widespread deforestation of mangroves in Vietnam triggered massive erosion (Mazda et al., 2002). To protect the coast from natural disasters like cyclone, many countries in South and Southeast Asia are involving themselves in restoration and preservation activities of coastal greenbelts. Favourably, coastal areas protected by mangroves face less damage than non-mangrove barren areas (Ali, 1996). Plant species found in the mangrove are more resilient to natural disasters than non-mangrove species (Saenger & Siddiqi, 1993).

Hence, it is evident that responses of mangroves against the impact of climate change act as natural shields that protect humans from several damages and vulnerabilities (Fig. 8.2).

Sundarbans: A Natural Wall

For decades, the Sundarbans has been protecting the ecological balance of both Bangladesh and India. This forest is a composite and the largest single tract mangrove forest in the world. Spanning around 4,000 square miles on India–Bangladesh border, it stands as a green wall, mitigating losses even from the worst cyclones (Schwartzstein, 2019). Plenty of floral and faunal species form the natural ecosystems of this forest. Millions of people from both Bangladesh and India depend largely on this mangrove forest for their livelihood. During FGD, one of the TRUs, Abdus Salam, said

> We are like the son of the forest. It fosters us, feeds us, protects us. We cannot survive without the existence of the Sundarbans.

The TRUs gather resources from both coastal and inland belts of the forest. Apart from socio-economic benefits, it has a striking influence on the existence and natural harmony of humans along with other species. It is regarded as a shield or natural wall as it protects the people and environment from natural disasters such as storms and cyclones. Suffering from the colossal winds itself is mitigated so that the speed weakens and other regions bear less loss. Anju-Ara, a female TRU whose house is adjacent to the forest, said

> When we hear the weather forecast calling people to go nearby shelter since a severe cyclone is going to hit the region, we generally go to the shelter leaving our house and all other belongings behind. We often think we are going to lose everything. It is astonishing, however, that when returning home, we often see that nothing is lost. The cyclone goes over the forest causing no or little harm to us. The Sundarbans takes all the harshness of the cyclone. It works as the protection to us.

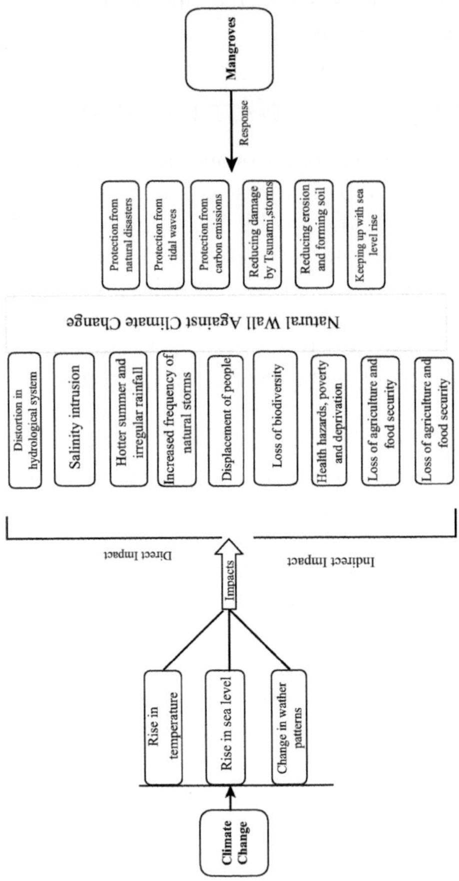

Fig. 8.2 Mangroves as a natural wall to climate change (*Source* Prepared by authors)

Table 8.3 Economic value of the ecosystem services

Services	*Economic value (US$ '000 per year)*
Regulatory (Coastal protection)	42,250
Provisioning	744
Cultural	42
Total	43,034

Source Uddin (2013)

To safeguard the people and coast from cyclones, the forest endures enormous losses. Another TRU, Shahjahan, said

> Though Sundarbans works as a protection against cyclones, salinity and tsunami, it faces plenty of damages. Trees and other vegetation get uprooted, rivers and canals become full of leaves and branches which cause water pollution. It, therefore, causes harm to aquatic lives too.

Besides protecting the environment from natural calamities, it also plays a vital role in preventing salinity intrusion, storing emitted in nature, reducing the height of the tidal surge, surviving against sea-level rise and so on. It halts the salinity of the land, keeps environmental balance by preventing pollution and also acts as a watchman against natural disasters (Mamun, 2019). The economic value of regulating services as a form of coastal protection by the forest is estimated to be almost US$ 42 million/year which is the highest among all ecosystem services (Table 8.3).

This estimation implies that the forest works as the 'coast guard' in Bangladesh against natural hazards. The economic value of this is much higher than what is estimated if all other tangible and intangible services are included (Uddin, 2013). The TRUs regard this forest as a part of their lives. Therefore, its true value cannot be estimated using the market price regime (see chapter 6 & 7). The following section categorically explores the role of mangroves to fight climate change.

Protection from Natural Disasters

The forest acts as a coastal greenery shield. For the southwestern coast of Bangladesh, it has been playing this protective role in a natural manner for a long time. Cyclones in the forest with a speed of 100 km per hour lose power while crossing the forest (Mamun, 2019). After crossing through the forest, the storm becomes weak before hitting the locality. Likewise, the tidal wave is also obstructed by the forest, before affecting the coastal area. As a result, severe damages are reduced.

Mangroves protect the mainland from cyclones and tsunami (Danda, 2019). Most of the major cyclones hitting the coast of Bangladesh are of severe types (Table 8.3). These cyclones cause enormous losses each year (see Titumir & Paran, 2022). The harshness and severity of these cyclones are much more significant than the number of cyclones occurring. About 53% of the world's deaths due to cyclones occur in Bangladesh and about 23% in India (Rahman & Rahman, 2015). However, it is worthy to note that these two countries are hit by only 4.27% of the world's storms (see frequency in Fig. 8.3).

This mangrove forest saved Bangladesh from several storms and cyclones like Sidr, Aila and Bulbul which could have been the worst and more life-taking. The strong tidal wave got disrupted in the dense mangrove area and the strong wind got weaker by getting impeded in the branches of the dense forest. Losses and damages got largely reduced, due to the forest working as a speed-bump. The wind speed of cyclone Bulbul,

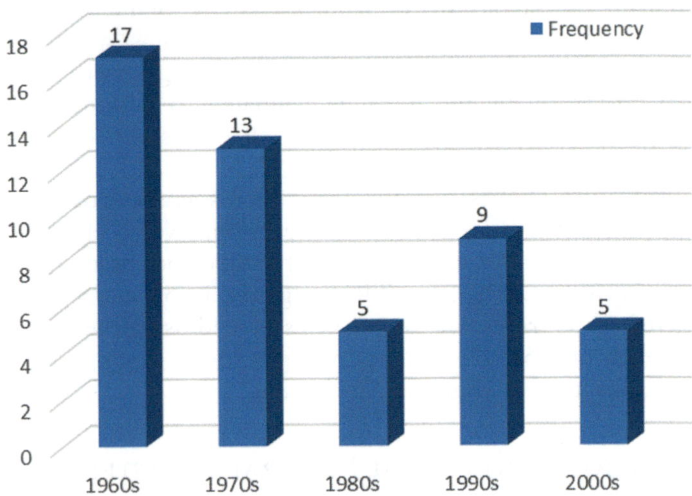

Fig. 8.3 Frequency of cyclones hitting Bangladesh during 1960s to 2000s (50 years) (*Source* Saha and Khan [2014])

ranged between 115 and 125 kilometres per hour which was lowered to approximately 93 km per hour when it reached the Bangladesh coastal belts (Roy, 2019). As a result, it hit Khulna, an administrative division in south-western region, with the least force.

Frequent tropical cyclones hitting the coast of Bangladesh causes serious damage to the forest (Table 8.4). The forest works as shield to the coast bearing heavy loss and damage itself. For example, the super cyclone Amphan that hit the coast of Bangladesh and India on 20 May 2020 uprooted around 10,500 *Garan* trees (*Ceriops decandra*) along with some other species with an estimated cost of Tk 2.25 crore in the Bangladesh[2] part of the forest.

Protection from Tidal Waves

Mangroves reduce the power of wind, therefore, stretching the height of waves passing through the waterways. This decreases the sediment eroding capability and structures such as dikes and sea walls damaging the strength of the waves. When waves enter the forest during rising tides, they get tangled in roots and branches, for which they instantly lose their height by between 13 and 66% over 100 m of mangroves (Spalding et al., 2014). Reducing winds across the surface of water bodies bears the regeneration of waves. The denser the forest, the more obstacles it makes. In this way, the forest works as a saviour for both marine coastal and terrestrial regions, protecting the country from high tidal waves that could otherwise cause flash floods.

Protection from Carbon Emission

Mangroves absorb carbon. They can protect the natural environment from pollution and calamities with their biological formation and functions. Falling leaves and branches turn into sediments and release nutrients causing bacterial function. The organic peat forms carbon-rich soils. Though it varies from location to location, the major share of carbon is stored in the soils, which makes this ecosystem a significant actor in the longer-term mitigation of carbon from the atmosphere. Without

[2] See the report https://tbsnews.net/environment/amphan-causes-tk225cr-damage-sundarbans-85654?fbclid=IwAR3FXiZCb_mwoSPipXP3LUebspcTPiMiRkjMV5c-FtjefM6BbUzxACmlOhI (last accessed 28 February 2020).

Table 8.4 Types of various major cyclones hitting Bangladesh over the years

Date	*Types of cyclones*	*Landfall area*	*Maximum wind speed in km/hr*	*Tidal surge height in ft*
23.10.1970	Severe Cyclonic Storm of Hurricane intensity	Khulna-Barisal	163	–
12.11.1970	Severe Cyclonic Storm with a core of hurricane wind	Chittagong	224	10–13
28.11.1974	Severe Cyclonic Storm	Cox's Bazar	163	9–17
10.2.1981	Cyclonic Storm	Khulna	120	7–15
15.10.1983	Cyclonic Storm	Chittagong	93	–
09.11.1983	Severe Cyclonic Storm	Cox's Bazar	136	5
24.04.1985	Severe Cyclonic Storm	Chittagong	154	15
29.11.1988	Severe Cyclonic Storm with a core of hurricane wind	Khulna	160	2–14.5
18.12.1990	Cyclonic Storm (crossed as a depression)	Cox's Bazar Coast	115	5–7
29.04.1991	Severe Cyclonic Storm with a core of hurricane wind	Chittagong	225	12–22
02.05.1994	Severe Cyclonic Storm with a core of hurricane wind	Cox's Bazar-Teknaf Coast	220	5–6
25.11.1995	Severe Cyclonic Storm	Cox's Bazar	140	10
19.05.1997	Severe Cyclonic Storm with a core of hurricane wind	Sitakundu	232	15
27.09.1997	Severe Cyclonic Storm with a core of hurricane wind	Sitakundu	150	10–15
20.05.1998	Severe Cyclonic Storm with core of hurricane winds	Chittagong Coast near Sitakunda	173	3

(continued)

Table 8.4 (continued)

Date	*Types of cyclones*	*Landfall area*	*Maximum wind speed in km/hr*	*Tidal surge height in ft*
28.10.2000	Cyclonic Storm	Sundarban Coast near Mongla	83	–
12.11.2002	Cyclonic Storm	Sundarban Coast near Raimangal River	65–85	5–7
19.05.2004	Cyclonic Storm	Teknaf-Akyab Coast	65–90	2–4
15.11.2007	Severe Cyclonic Storm with core of hurricane winds (SIDR)	Khulna-Barisal Coast near Baleshwar river	223	15–20
25.05.2009	Cyclonic Storm (AILA)	West Bengal-Khulna Coast near Sagar Island	70–90	4–6
16.05.2013	Cyclonic Storm (MAHASEN)	Noakhali-Chittagong Coast	100	–
30.07.2015	Cyclonic Storm (KOMEN)	Chittagong-Cox's Bazar Coast	65	5–7
21.05.2016	Cyclonic Storm (ROANU)	Barisal-Chittagong Coast near Patenga	128	4–5
30.05.2017	Severe Cyclonic Storm (MORA)	Chittagong-Cox's Bazar Coast near Kutubdia	146	

Source BMD (2017)

Table 8.5 Result of 2010 carbon inventory for the Sundarbans

Carbon pool	*Carbon density (Mg/ha)*
Above ground trees, seedlings and saplings	83.8
Below ground (roots) of trees, seedlings and saplings	36.9
Nontree vegetation such as goran (Ceriops decandra), golpata (Nypa fruticans) and others	10.7
Downed wood	4.3
Soil (0–100 cm depth)	121.0
Total	256.7

Source Forest-Department (2011) cited in Mamun et al., (2017)

mangroves, such a huge share of stored carbon will be released back into the atmosphere. The Sundarbans functions effectively by storing carbon after absorbing them from nature. There have been numerous mills and factories set up around it. These activities are threatening the natural functioning of the forest. The establishment of the *Rampal* coal-based power plant will cause more pollution in the vicinity of the forest and will cause degradation of the forest's capability to alleviate carbon and other harmful emissions. The inventory of the carbon insulated by the forest, however, implies that of Bangladesh's annual fossil CO_2 emissions, the annual sequestration rate over the last 13 years is 4.8 Mt CO_2 per year, or ~10% (Mamun et al., 2017).

Protection from Erosion

The ever-changing waves and currents bring sediments to the coast but at the same time erode the soil. As mangroves reduce the force of waves and currents, they in turn also reduce erosion. The roots, layers of fallen leaves and branches trap the sediments and soil. Waterlogged soils of mangroves with a low amount of oxygen form a layer of peat that becomes thick over time (Spalding et al., 2014). The forest prevents coastal erosion in this manner. The strong roots and woody materials save the river banks from getting eroded. By causing increased sedimentation, the probability of erosion becomes diminutive for which the embankments become safe.

Keeping up with Sea-Level Rise

Mangrove soils usually grow by capturing riverine or coastal sediments that pass through the forest and form the roots, fallen leaves and branches into organic matters. The roots help trapping and binding sediments which push the level of soil upward. This soil formation mechanism allows mangroves to sustain their pace with the rising sea level that occurs due to global warming. Sometimes the layers of mangroves peat are 6 m deep or more on which the mangroves sit (Spalding et al., 2014).

Conclusions

This chapter explores the role of this forest as a natural wall. The forest has a salient influence on the existence and natural harmony of humans along with other species. It is being regarded as a shield to protects the people and environment from natural disasters such as storms and cyclones. It is a known fact that climate change is a serious concern and there are numerous anthropogenic drivers that cause degradation of the natural ecosystem which in turn leads to climate change. Hence, natural disasters and damages resulting from climate change are on the rise.

This impact of climate change on the forest is also acknowledged by the TRUs. A survey questionnaire was created regarding the impact of climate change on nature, alongside their life and livelihoods which was distributed among the TRUs to get their opinion on the matter. Just like in the previous chapter, a Likert scale was used here to analyse the findings. In response to this, most of the respondents agreed that the salinity in the coastal areas has been increasing, affecting both life and livelihoods. This was also decreasing the yield of crops coming from agricultural lands. The survey also found that the availability of fish in the rivers had decreased in number which was mostly due to over-extraction. Besides, PPGIS conducted for LULC prediction using the Maxent model suggests that climatic variables, geomorphological conditions, soil nutrients and anthropogenic factors have a significant contribution to biodiversity loss (big trees/forest coverage). In 2050, almost half of the Sundarbans in the study area will become mere fallow land.

The mangroves have always acted as a natural wall and have saved the country from numerous hazards. The forest has been offering an effective physical barrier against the impact of climate change and saving both lives and property. It also provides multiple benefits and services that are crucial to moving beyond the vulnerabilities that may result from climate change. These include soil surface dynamics in mangroves and the aerial roots of the mangrove trees. Therefore, causing harm towards this forest can create a negative impact on the ecological cycle of the forest's biodiversity. This will further affect human beings and other floral and faunal species in a fatal way.

Thus, mangrove forest is one of the best gifts to Bangladesh. The value of this mangrove forest cannot be calculated in monetary terms fully. It acts as a powerful shield against natural hazards. The significance of its services especially while providing protection from climate change is related to the sole existence of this land. Unfortunately, its role as a natural wall against natural disasters is often undervalued. Several anthropogenic pressures are causing severe damages to this valuable forest. If these cannot be checked, Bangladesh is going to lose its 'natural wall' that assists in its fight against climate change. The impacts of climate change are already evident. The natural ecosystem has been found to be effective, globally against these impacts. The conservation and restoration of it along with sustainable use and benefit-sharing (appropriation) between humans and nature is an important requirement. Moreover, any form of man-made harm to it is going to destroy its age-old capability to fight against natural hazards. Hence, it ought to be one of the pressing responsibilities, to preserve this natural wall.

References

Ali, A. (1996). Vulnerability of Bangladesh to climate change and sea level rise through tropical cyclones and storm surges. *Water, Air, and Soil Pollution, 92*, 171–179.

Badola, R., & Hussain, S. A. (2005). Valuing ecosystem functions: An empirical study on the storm protection function of Bhitarkanika mangrove ecosystem, India. *Environmental Conservation, 32*(1), 85–92.

BMD. (2017). *List of major cyclonic storms from 1960 to 2017*. Bangladesh Meteorological Department, Government of the People's Republic of Bangladesh, Dhaka, Bangladesh. http://live3.bmd.gov.bd/p/Historical-Cyclones/. Last accessed on 23 March 2021.

Barbier, E. B. (2016). Sustainability and development. *Annual Review of Resource Economics, 8*(1), 261–280.

Brander, L. M., Wagtendonk, A. J., Hussain, S. S., McVittie, A., Verburg, P. H., de Groot, R. S., & van der Ploeg, S. (2012). Ecosystem service values for mangroves in Southeast Asia: A meta-analysis and value transfer application. *Ecosystem Services, 1*(1), 62–69. https://doi.org/10.1016/j.ecoser.2012.06.003

Chowdhury, A., & Brahma, A. (2019). Environment and wellbeing: Eco-psychiatry in Sundarban Delta, India. *International Journal of Management and Social Sciences, 14*(2), 37–53.

CCC. (2008). *Climate change, gender and vulnerable groups in Bangladesh.* Climate Change Cell, Department of Environment, Ministry of Environment, Forest and Climate Change, and Component 4b, Comprehensive Disaster Management Programme, Ministry of Food and Disaster Management, Dhaka, Bangladesh.

Danda, A. A. (2019). *Environmental security in the Sundarbans in the current climate change era: Strengthening India-Bangladesh Cooperation* (ORF Occasional Paper, Observer Research Foundation).

Das, S., & Vincent, J. R. (2009). Mangroves protected villages and reduced death toll during indian super cyclone. *Proceedings of the National Academy of Sciences of the United States of America, 106*(18), 7357–7360. https://doi.org/10.1073/pnas.0810440106

Elith, J., Phillips, S. J., Hastie, T., Dudík, M., Chee, Y. E., & Yates, C. J. (2011). A statistical explanation of maxent for ecologists. *Divers and Distributions, 17*, 43–57. https://doi.org/10.1111/j.1472-4642.2010.00725.x

FAO. (2007). *The world's mangroves 1980–2005* (FAO Forestry Paper 153). FAO.

Flaherty, M., Szuster, B., & Miller, P. (2000). Low salinity inland shrimp farming in Thailand. *Ambio, 29*, 174–179.

Flaherty, M., & Vandergeest, P. (1998). "Low-salt" shrimp aquaculture in Thailand: Goodbye coastline, Hello Khon Kaen! *Environmental Management, 22*, 817–830. https://doi.org/10.1007/s002679900150

Forest Department. (2011). *Project concept note, collaborative REDD$_+$IFM Sundarbans Project (CRISP)*. Ministry of Environment and Forests.

Government of Bangladesh. (2008). *Bangladesh climate change strategy and action plan*. Government of Bangladesh. http://www.fao.org/faolex/results/details/en/c/LEX-FAOC163539/#:~:text=Bangladesh%20Climate%20Change%20Strategy%20and,the%20challenge%20of%20climate%20change. Last accessed on 20 March 2021.

Granek, E. F., & Ruttenberg, B. I. (2008). Changes in biotic and abiotic processes following mangrove clearing. *Estuarine Coastal and Shelf Science, 80*(4), 555–562.

Krauss, K. W., McKee, K. L., Lovelock, C. E., Cahoon, D. R., Saintilan, N., Reef, R., & Chen, L. (2014). How mangrove forests adjust to rising sea level. *New Phycologist, 202*(1), 19–34.

Mamun, H. R. (2019). Sundarbans: The natural shield of Bangladesh. https://dailyasianage.com/news/207625/sundarbans-the-natural-shield-of-bangladesh. Last accessed on 13 March 2021.

Mamun, M. M. A. A., Masum, K. M., Sarker, A. H. M. R., & Mansor, A. (2017). Ecosystem services assessment using a valuation framework for the Bangladesh Sundarbans: Livelihood contribution and degradation analysis. *Journal of Forestry Research, 28*, 1–13. https://doi.org/10.1007/s11676-016-0275-5

Marcer, A., Saez, L., Molowny-Horas, R., Pons, X., & Pino, J. (2013). Using species distribution modelling to disentangle realised versus potential distributions for rare species conservation. *Biological Conservation, 166*, 221–230.

Marois, D., & Mitsch, W. J. (2015). Coastal protection from Tsunamis and cyclones provided by mangrove wetlands: A Review. *International Journal of Biodiversity Science, Ecosystem Services & Management, 11*(1), 71–83.

Mazda, Y., Magi, M., Nanao, H., Kogo, M., Miyagi, T., Kanazawa, N., & Kobashi, D. (2002). Coastal erosion due to long-term human impact on mangrove forests. *Wetlands Ecology and Management, 10*, 1–9.

Phillips, S. J. (2017). *A brief tutorial on Maxent.* http://biodiversityinformatics.amnh.org/open_source/maxent/. Retrieved 20 October 2020.

Phillips, S. J., Andersonb, R. P., & Schapire, R. E. (2006). Maximum entropy modeling of species geographic distributions. *Ecological Modelling, 190*, 231–259. https://doi.org/10.1016/j.ecolmodel.2005.03.026

Rahman, M. A., & Rahman, S. (2015). Natural and traditional defense mechanisms to reduce climate risks in coastal zones of Bangladesh. *Elsevier, 7*, 84–95.

Rahman, M. M., Rahman, M. M., & Islam, K. S. (2010). The Causes of deterioration of Sundarbans mangrove forest ecosystem of Bangladesh: Conservation and sustainable management issues. *Aquaculture, Aquarium, Conservation & Legislation International Journal of the Bioflux Society, 3*(2), 77–90.

Roy, R. (2019). Sundarbans: A natural shield against natural disasters. http://m.theindependentbd.com/home/printnews/224287. Last accessed on 13 March 2021.

Saenger, P., & Siddiqi, N. A. (1993). *Land from the sea: The mangrove afforestation program of Bangladesh.* Southern Cross University.

Saha, M., & Khan, A. N. (2014). Changing profile of cyclones in the context of climate change and adaptation strategies in Bangladesh. *Journal of Bangladesh Institute of Planners, 7*, 63–78.

Schwartzstein, P. (2019). This vanishing forest protects the coasts—And lives—Of two countries: Rising waters and illicit logging are killing the trees in the Sundarbans, the natural wall that protects the India-Bangladesh Coast. *National Geographic Magazine.* https://www.nationalgeographic.com/magazine/2019/07/sundarbans-mangrove-forest-in-bangladesh-india-threatened-by-rising-waters-illegal-logging/. Last accessed on 13 March 2021.

Spalding, M., McIvor, A., Tonneijck, F. H., Tol, S., & van Eijk, P. (2014). Mangroves for coastal defence. Guidelines for coastal managers & policy makers. *Wetlands International and The Nature Conservancy.*

Titumir, R. A. M., & Paran, M. S. (2022). Human-nature cooperation for well-being: Community understanding on one health approach in the COVID-19 era in the Sundarbans. In UNU-IAS & IGES (Eds.), *Nexus among biodiversity, health, and sustainable development in managing socio-ecological production landscapes and seascapes (SEPLS)* (Satoyama Initiative Thematic Review, Vol. 7). United Nations University Institute for the Advanced Study of Sustainability.

Uddin, S. M. (2013). *Economic valuation of Sundarbans mangrove ecosystems-Bangladesh.* LAP Lambert Academic Publishing.

Unnayan Onneshan. (2020). *Biodiversity, climate change and traditional resource users in the Sundarbans: An exploration through Public Participation Geographic Information System (PPGIS)*. Unpublished Research Report.

Woodruff, J., Camargo, S. J., & Irish, J. L. (2013). Coastal flooding by tropical cyclones and sea-level rise. *Nature, 504*(7478), 44–52.

CHAPTER 9

Biodiverse Adaptation to Climate Change: Community-Based Mangrove Agro Aqua Silvi (CMAAS) Culture

Rashed Al Mahmud Titumir and Tanjila Afrin

Introduction

This chapter explores biodiversity adaptation options in response to climate change, and in this regard, introduces a new method, which is known as Community-based Mangrove Agro Aqua Silvi (CMAAS) culture. This is practised in Bangladesh's coastal region, notably in certain

R. A. M. Titumir (✉)
Department of Development Studies, University of Dhaka, Dhaka, Bangladesh
e-mail: rt@du.ac.bd; rtitumir@unnayan.org

Unnayan Onneshan, Dhaka, Bangladesh

T. Afrin
Department of Development Studies, Bangladesh
University of Professionals, Dhaka, Bangladesh
e-mail: tanjila.afrin@bup.edu.bd

R. A. M. Titumir (ed.), *Sundarbans and its Ecosystem Services*, Sustainable Development Goals Series,
https://doi.org/10.1007/978-981-19-3000-3_9

areas of Khulna, Satkhira and Bagerhat districts, where a section of population that is prone to natural disasters is trying to harvest few floral and faunal species found in the forest and on lands that were previously used for other purposes. This practice is a substitute for Commercial Shrimp (CS) culture. The chapter represents the overall picture of the CMAAS culture in relation to the impacts of climate change in the coastal region of Bangladesh and to the conservation of the Sundarbans mangrove ecosystem.

The chapter opens by giving a brief overview of CMAAS culture and how it came into the picture. While explaining the same, it takes into account the numerous factors related to it such as climate change patterns, natural disasters, intensive shrimp cultivation and salinity intrusion. It goes on to present a schematic diagram to explain the aforementioned factors along with their drivers and impacts and how they are linked to the emerging CMAAS culture. The culture incorporating cultivation methods of the mangroves practised within the forest are also mentioned here. It is supported by another schematic diagram of the general farming procedure of mangroves. Furthermore, it includes the CMAAS culture's incorporation of the cultivation methods for aquatic species that is backed by another diagram. The cultivation methods of both mangrove species and aquatic species include many sub-factors such as site selection, species selection, farm construction, seed collection, maintenance, economic costs and so on. The chapter explains each of these points in context and how they contribute to cultivation under this culture. The provisioning uses are also mentioned as to which forest resource is used for what purpose. Finally, a comparative analysis has been done between CMAAS culture and Commercial Shrimp (CS) culture.

The Emergence of CMAAS Culture: Background Reasons

The coastal area of Bangladesh comprises around 32% of the country (Parvin et al., 2016). This vulnerable coastal zone encompasses around 19 of Bangladesh's 64 districts, home to 30% of the country's population and more than half of whom are impoverished (Parvin et al., 2009). The majority of the coastal regions are low-lying, rising approximately one metre above the average sea level. As a consequence, at high tides, the lowlands are usually submerged by seawater. Coastal Bangladesh comprises a wide range of ecosystems and biological resources. It is also the home to 36 million people approximately. The livelihood options of most households are agriculture-based along with activities such as fishery, forestry, shore transportation and salt panning. The mangrove forest acts as the main resource system which is also known as the natural power house. It lies along the south-western belt of the coast (Fig. 9.1).

Over the last five decades, usage of coastal land in Bangladesh has faced major changes. Natural calamities such as cyclones and tidal floods, followed by climate change, salinity intrusion, large-scale polder development and intensive shrimp farming, have transformed the environment of the country's whole coastline area since 1950. Because of these changes in coastal land usage pattern, the coastal community's agriculture, crop production, food and water supply and livelihood have all suffered severe consequences. In response, local communities have been practising different production methods to adapt to new situations. CMAAS culture is one of those methods, which is developed by them. It is based on their own knowledge that adds to their resilience ability in the face of climate change. Here, it is necessary to specify about the background reasons for the emergence of this culture before proceeding into details.

Climate Change, Natural Disasters and Salinity Intrusion

Not every part of Bangladesh is vulnerable to climate change in the same manner. The area that is most vulnerable is the coastal area. Coasts are widely thought by climate scientists to be increasingly vulnerable to the growing hazards of catastrophic climatic events. Climate change is expected to increase the frequency and severity of climatic risks and disasters. People in the region are finding it challenging to preserve their lives and livelihoods due to climate change resulting in newer and severe natural catastrophes as well as saline intrusion. Hence, the people residing in these areas seek for alternative livelihood options and strive to adapt to the changes.

Intensive Shrimp Cultivation and Salinity Intrusion

The most serious anthropogenic danger to the mangrove forest is the conversion of lands into Commercial Shrimp (CS) farms. The growth in the number of the farms is mostly the result of quasi-legal action. The farms are run by wealthy local stakeholders. This is precisely done by the rich fishermen (who are not indigenous people), who are politically connected and enjoy the favour of the administrative authorities at both

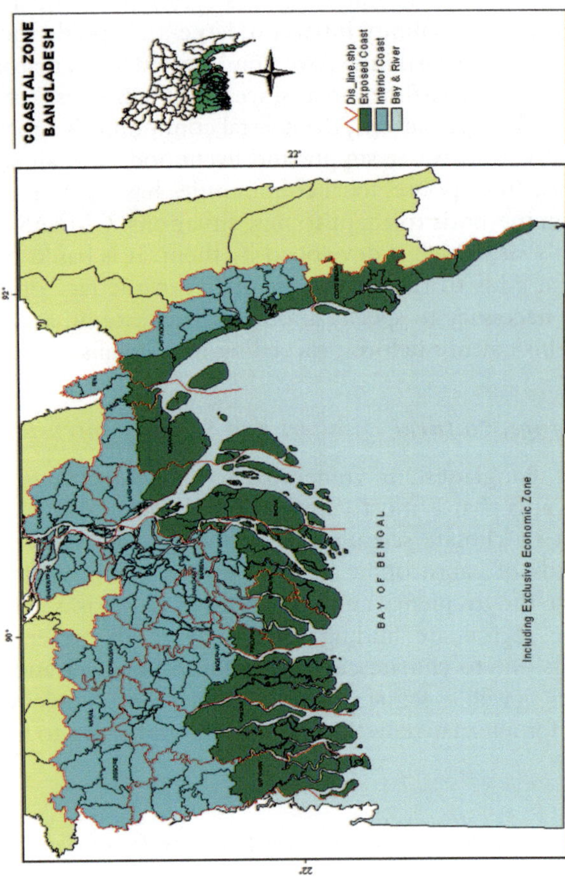

Fig. 9.1 Map of the coastal zone of Bangladesh (*Source* ICZMP [2005] as cited in Rezaie [2015])

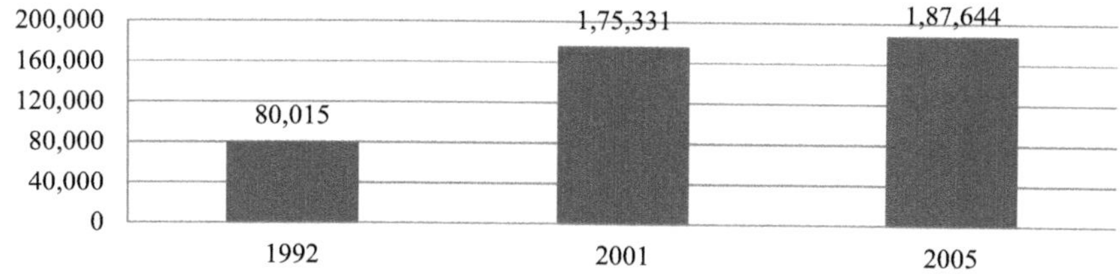

Fig. 9.2 *Bagda* shrimp cultivated areas adjacent to the Sundarbans (in hectares) (*Source* Authors' calculation based on IUCN [2014])

the local and national levels. During 1992 to 2005, there has been an increase in the cultivation of shrimp (Bagda–*Penaeus monodon*) in regions next to the forest (in hectares) (Fig. 9.2). Shrimp pond building degrades and destroys mangrove habitat in a variety of ways. A shrimp farming pond, for example, reaches the end of its productive life three to six years after construction and is subsequently abandoned.

As a result, the cultivators have to move along the shore, destroying mangroves in the process, in order to establish more new ponds. This type of exploitative shrimp farming raises the salinity of the soil and changes the composition of the soil in the region. Indeed, severe salinisation is already wreaking havoc on Bangladesh's south-west coastal area, particularly between October and May. Salinity levels in the region have risen over time, according to laboratory tests of water and soil samples (World Bank, 2016). For instance, the extent of aquatic salinity intrusion can be visualised in the following Fig. 9.3. The salinity has been increasing because of the rise in sea level and intensive shrimp cultivation. Therefore, there emerges the need to search for alternative farming practices.

Polders, Embankments and Dam Construction

The construction of large polders and embankments result in the loss of smaller tidal flow to the hinterland which not only affects natural hydrologic systems but also hampers tidal brackish and monsoonal freshwater tides. Two direct consequences of this are negative sediment ratio that cannot sufficiently cover up for land subsidence anymore, and increased salt concentrations in the surface layers (Falk, 2015). Large upstream dams such as the Indian Farakka Barrage, built in 1974, has decreased

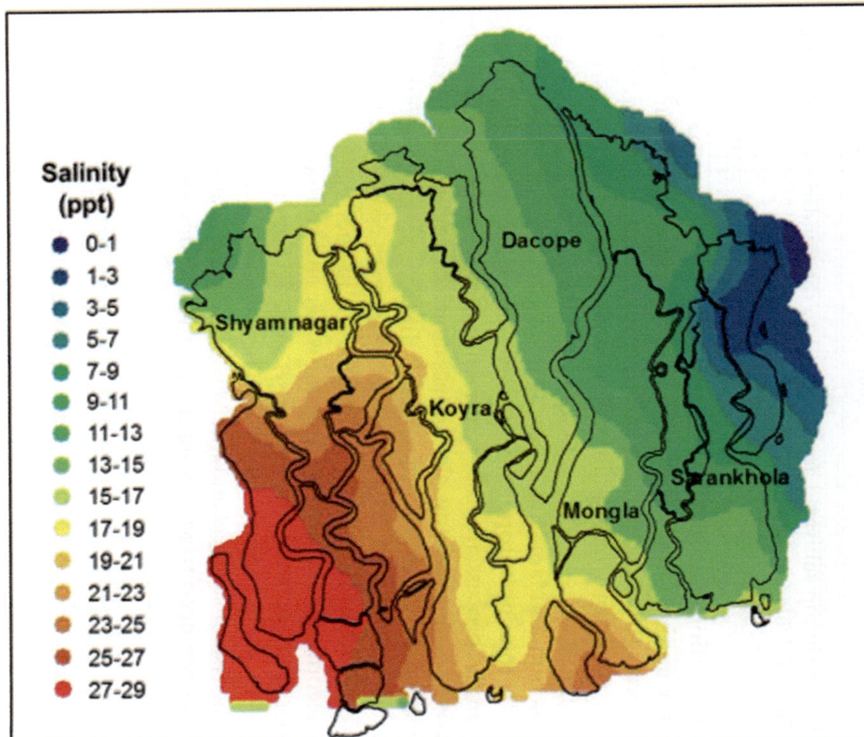

Fig. 9.3 Estimated aquatic salinity in the Sundarbans adjacent regions (*Source* Dasgupta et al. [2017])

river discharge or the freshwater flow in the downstream rivers—especially during the dry season. Islam and Gnauck (2009) showed that 'the Ganges flow was 3700 m3/s in 1962 whereas it was reduced to only 364 m3/s in 2006'. Figure 9.4 shows the decrease in Ganges water flow and salinity intrusion. It reveals that after the construction of Farakka dam, water discharge in the downstream rivers has decreased while the salinity of water has increased. These can be seen by the downward and upward sloping curves portrayed in the figure.

Sarker et al. (2016) mentioned in another report that Ganges' freshwater flow into the forest has declined from 3700 $m^3 s^{-1}$ to364 $m^3 s^{-1}$ since the construction of the Farakka dam. This has increased siltation and salinity inside the mangrove forest and in the coastal area of Bangladesh.

In this way, the total impact of natural and human-induced hazards and disasters are threatening both the social and the physical environment of the coastal zone in multiple ways. It has critically affected almost every aspect of life in the coastal zone. Agriculture, the most significant livelihood option of the coastal people, is the most vulnerable as it is negatively impacted by salinity intrusion. Hundreds acres of arable land fall victim to very high soil salinity every year and consequently is rendered sterile. Furthermore, the saline-induced lands cannot be brought under any other productive use. Virtually, nothing remains but fallow lands. It is inferred that agricultural vulnerability will be further provoked, by climate change coupled with extreme population density. Currently, reliability on crop culture among the coastal people is declining which is pushing them towards a crisis.

To comply with the emergent adverse conditions, and utilise the saline rich croplands, the local communities have immediately promoted a number of intriguing adaptation measures by applying their creative ideas and traditional knowledge. They include *mele* cultivation, crab culture, CMAAS culture, oyster culture, double-decker crop cultivation and so on (see Baten & Kumar, 2010; Kabir & Hossian, 2008;Unnayan Onneshan, 2010; Titumir, 2011, 2015). This chapter particularly focuses on the CMAAS culture. The practice of this mixed cultivation method began because of the curiosity of a few individuals with an inventive mind, and it was carried out on a modest scale.

Figure 9.5 shows the natural and anthropogenic drivers, effects and reactions to climate change and salinity intrusion in the coastal part of Bangladesh as well as their connections to the emergence of different types of cultivation methods including CMAAS culture. It can be seen

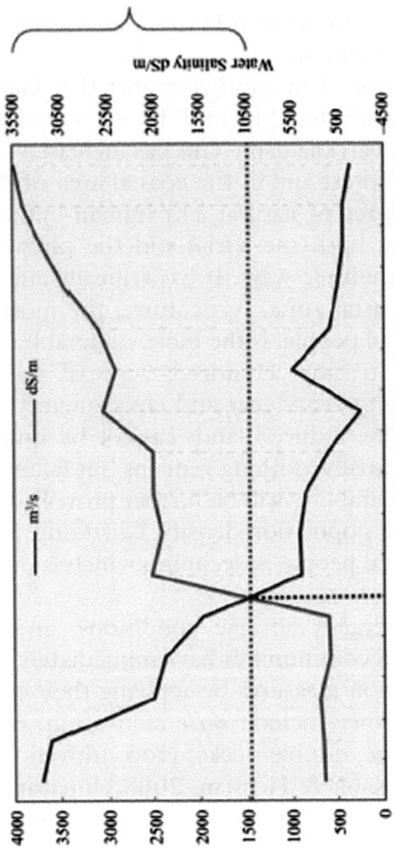

Fig. 9.4 Ganges water flow decrease and salinity intrusion (*Note* Vertical dotted line indicates the construction of the Farakka dam in 1975 (*Source* Islam and Gnauck [2009])

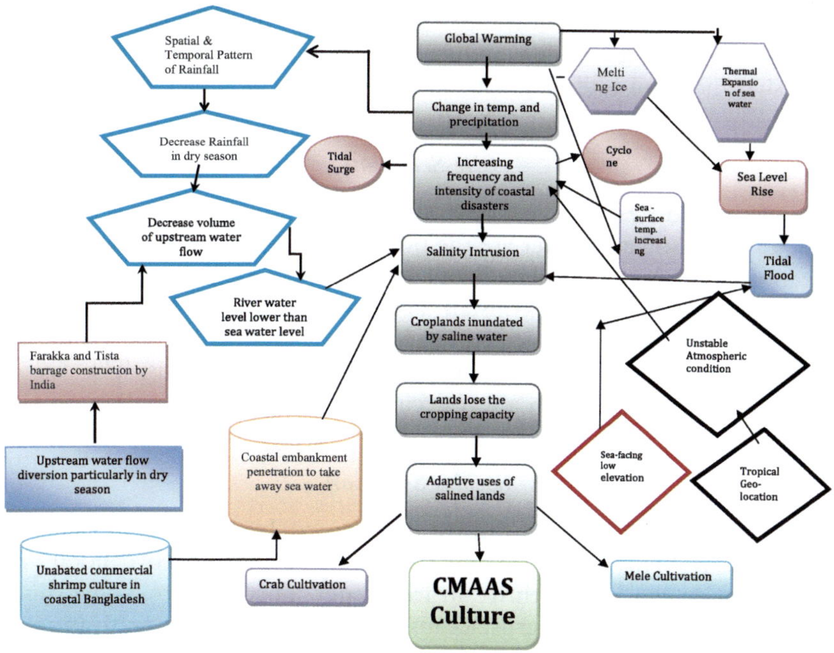

Fig. 9.5 Schematic diagram representing natural and anthropogenic drivers, impacts of/and responses to climate change and salinity intrusion at coastal Bangladesh and their linkages to the emergence of CMAAS culture (*Source* Adapted from data reservoir of Unnayan Onneshan)

how the global warming has branched out to various issues, each of which has, in turn, caused some problems or the other that has directly and indirectly hampered the whole ecosystem. Altogether, these issues in some form or the other have led to the emergence of the CMAAS culture.

Cultivation Method of Mangroves Under CMAAS Culture

Mangroves generally grow in the intertidal areas. For this, curiosity arises as to how they are grown by the community people in freshwater or brackish water and swampy lands. Despite CMAAS culture being an integrated cultivation of mangroves and aquatic species on the same piece of swampy land, their cultivation methods are independently brought up here for a better understanding. The mode of cultivation is largely extensive and semi-intensive with low capital inputs.

Site Selection

One of the most essential aspects of CMAAS culture is site selection for farm growth. Soil and water are the primary concerns throughout the site selection procedure. Farmers favour places with yielding, mud-spattered and waterlogged soils and constant access to brackish or freshwater. The majority of the farms are located near coastal rivers, which are always full of water and also draw water from them on a regular basis. The size of cultivation sites varies from a few decimals to a few *bighas*.

Species Selection

The selection of species is mostly determined by the water quality at the farm site. This is due to the fact that not all mangrove floral and faunal species can live or thrive in freshwater. Furthermore, some freshwater fish, mollusks and crustaceans are not suitable for cultivation in brackish water farms. Generally, it is the practice of integrated cultivation of some mangrove faunal species—crabs, oyster or fishes (e.g. shrimps and Bhetki [*Lates calcarifer*]) and floral species—Golpata (*Nypa fruticans*), Keora (*Sonneratia apetala*), Goran (*Ceriops decandra*), Hargoza (*Alanthus ilicifolius*) and Baen (*Avicennia species*), etc. at the same time on any swampy land of brackish water. In addition, integrated cultivation of some mangrove floral species like Golpata and a few faunal species like Tengra (*Mystus tengara*), Baila (*Awaous guamensis*), Tilapia (*Tilapia nilotica*), etc. are practised in a fresh water swampy land.

Farm Construction

Following site and species selection, farm construction is the next phase. The farm is made up of marshy ground with 1–1.5 feet of water depth, which is bounded by a dyke that is 0.5–1.5 feet above the water level. An unexcavated central platform typically remains waterlogged. During the dry season, the platform is occasionally exposed. On these platforms, mangroves such as Golpata and Keora are planted. Furthermore, there is a canal of about 2–2.5 feet deep that runs along the farm dykes and is used

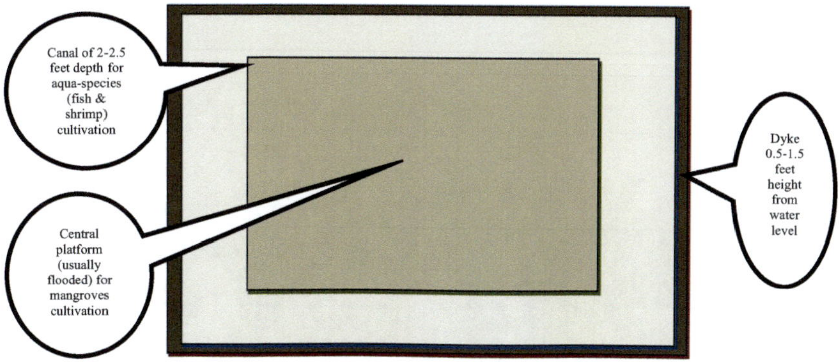

Fig. 9.6 Model of CMAAS culture farm (*Source* Adapted from data reservoir of Unnayan Onneshan)

to culture fish, shrimp and crabs. The farms normally have enough water to last the entire year and do not require water in the farming ponds. Figure 9.6 illustrates this, which is further portrayed in the real context (Fig. 9.7).

Cultivation of Mangrove Trees

The CMAAS culture farms are home to a handful of mangrove plants. These include Goalpata, Goran, Keora, Hargoza and Baen. The most frequent and dominating mangrove plant species grown is Golpata, which is followed by Keora. This is due to Golpata's greater adaptability, which allows it to thrive in freshwater swamps with little to no care and upkeep.

Seed Collection

The majority of the farm owners obtain mangrove seedlings from neighbouring riverbanks or places adjacent to the Khulna Port. The seeds, after dropping from the mangroves in the Sundarbans, enter into the coastal rivers and then it is carried by the tide. The seeds have the opportunity to settle down on river banks or get carried into brackish water shallows and are lodged into muddy bottoms. Without delay, they grow roots to take hold of the soil and accordingly stems grow to produce leaves. Any party, who is eager, can collect the seedlings without permission from the

Fig. 9.7 A CMAAS farm (Photo © Loban Rahman, Unnayan Onneshan)

authority. Others collect mangrove seeds floating in river water carried by the tide, while the remaining farm owners harvest seeds from the forest when extracting mangrove resources. Afterwards, they grow saplings from the seeds in nurseries at homesteads. Seed collection season of every kind of mangroves cultivated in the CMAAS culture farms is almost the same. All necessary mangrove seeds must be harvested between July and August.

Nursery Establishment and Management

A mangrove nursery is a place where mangrove seedlings are grown with extra care till they are ready to be planted in the farms. It helps develop good quality seedlings of optimum quantity at the right time. Seedbeds are prepared right after the collection of seeds from any source. The study reveals that yearly the seedbeds are prepared for all types of mangroves from July to August.

Maintenance of the Mangrove Plant Seedlings

The mangrove seedlings require some maintenance activities for their uninterrupted development, especially at the nursery stage.

Fencing
The nursery should be well fenced so that animals, wild or domesticated, cannot nip the seeds in the bud and harm the seedlings. Bamboo is commonly used as fences in the nursery.

Watering
Seeds and seedlings essentially require watering on a daily basis. Even before introduction in hardening beds, seeds need the use of tap or brackish water. But for hardened seedlings, brackish water must be used to adapt them to the field setting. Watering the plants every day early in the morning is ideal. Failure to water even for a day may have adverse effects on its growth and can even threaten its survival.

Weeding Out and Inspection
The seedbeds should be weeded at regular intervals to keep the seedling safe from weeds. Daily inspection should be done to ensure protection from insects and pests.

Plantation in the Farms

The seedlings develop to a minimum height of 1.5–2 feet in the nursery over a period of 1–1.5 months before being planted in the farm's central area. August is the beginning of the main planting season, which lasts through September. The density of mangrove trees planted in the platform varies on every farm, depending on whether the farmers favour mangrove or fish production. Because plantation density influences litter generation and the development of organic load, the density of mangrove trees has an impact on the intensity of aquaculture in the farm. It eventually affects the production and diversity of non-mangrove microfloral and faunal species. It ultimately causes an influence on the production and diversity of non-mangrove microfloral and faunal species. This may cover the lion's share of the diet of fish and other species cultured on the farm. In addition, mangrove density may vary based on the aquatic species to be cultured.

Farmers may also go for less dense mangroves (for example, one tree per sq. meter) for white fish culture for varieties like Telapia, Vetki, Tengra and Carps. Again, mangroves' density does not matter for shrimp and crab culture as these species are better reared in the shelter generated by the mangrove farms.

Care and Maintenance of the Mangroves

It requires little care and maintenance cost is negligible. Farm water exchange at a particular time is not essential for mangroves. They can survive both in fresh and brackish water and swampy lands. The biggest issue with mangrove farming is that they are occasionally attacked by viral diseases. Insecticide and pesticide usage can assist to safeguard them. They are cared for until they are ready to be harvested. Care continues until they are ready to be harvested. It is necessary to keep an eye on the mangroves if the farms are isolated.

Harvesting

The nursery development and plantation seasonality of the mangroves cultivated in the CMAAS culture farm are almost the same. Again, they widely vary from each other in terms of their harvesting period. Golpata and Goran possess the shortest harvest period. These can be collected and processed for market sale after 13–14 months of cultivation. Except for these two, all others need at least five years to mature before they can be harvested, with Possur taking the longest, at over eight years. Harvesting season for Golpata, the sole dominating mangrove plant in the CMAAS culture farm, is January through February, with new shoots beginning to sprout in March. Golpata is cut at a 45° angle.

Table 9.1 shows the general farming pattern of mangrove cultivation. Here, the names of the mangroves are classified along with their suitable timings (in months) for seed collection and seed bedding.

Overall, Fig. 9.8 shows a schematic diagram of a general farming procedure of the mangroves in CMAAS culture. It shows the breakdown of two different kinds of seed collection that are floating seeds from the rivers and collecting directly from the forest. It shows how both these lead to the growth of seedlings in homestead nurseries. Then, through a few more steps how the whole process is carried out is depicted in the diagram.

Table 9.1 General farming pattern (temporal) of mangroves cultivation

Name of Mangrove	*Seed collection and seed bedding*	*Plantation*	*Time from seed bedding to harvesting (month)*
Golpata	July	August	13–14
Keora	August	September	62
Hargoza	July	August	62
Baen	August	September	62
Possur	July	August	98
Goran	August	September	14

Source Adapted from data reservoir of Unnayan Onneshan

Uses of Mangroves

Golpata, the dominant plant species of CMAAS culture, is one of the valuable non-wood mangrove plant species. Its uses are plenty and diverse. Its leaf, popularly known as the poor man's tin-sheet, is extensively used in roofing the thatched houses in coastal Bangladesh, containing prolonged durability of about 2–3 years. If preservative treatment is obtained, the thatching materials may prolong the durability by up to 10–15 years.

The dried-up petioles, leaves, stems and fruit residues of Golpata are great sources of firewood while the skin of fresh petioles can be used to make good quality ropes. It bears fruits that are edible and hence have monetary value in the market. Fresh fruits can be eaten and also preserved for longer hours. Golpata also contributes to important medicinal uses. Ash from Golpata is reported to be used as an analgesic against tooth and headache.

The dominant mangrove in CMAAS culture next to Golpata is Keora which possesses a number of provisioning uses. It is extensively used in house building and furniture. Other than the leaves, the stems and roots are usually used as fuelwood. Hargoza is only used as fuelwood. Possur and Baen are also woody mangroves that are widely used in house construction and furniture. Table 9.2 shows different provisioning uses of the mangroves cultivated in the CMAAS culture farms.

Economic Cost and Return

The most important thing in any investment or project is the ratio of economic cost and return. The ultimate decision in the continuation or

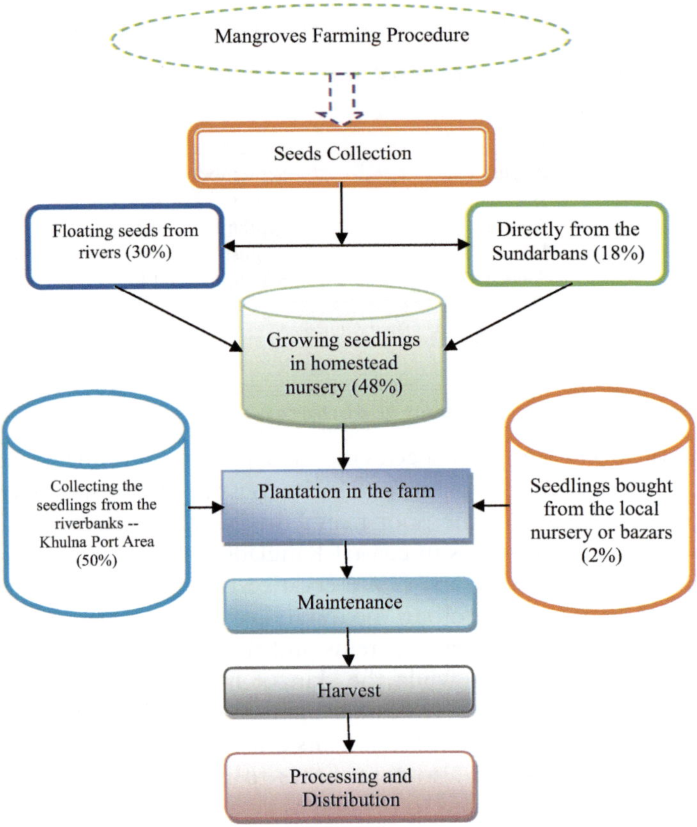

Fig. 9.8 A schematic diagram of general farming procedure of the mangroves in CMAAS culture (*Source* Adapted from data reservoir of Unnayan Onneshan)

discontinuation of any investment or project mostly depends on this. Community-based mangrove cultivation is found to be a very profitable coastal adaptation strategy. The annual average gross income from each *bigha* mangrove cultivation is around BDT 56250, while the whole cost is just approximately BDT 1800, resulting in a net gain of approximately BDT 54450 (Fig. 9.9). As a result, the cost–benefit ratio is 1:32.

However, it is not the case that all types of mangroves equally contribute to the gross income. The contribution varies from one mangrove species to another (Fig. 9.10). As an individual species, the

Table 9.2 Provisioning uses of the mangroves cultivated in the CMAAS culture farms

Mangrove	*Provisioning uses*	
Golpata	– House roofing – Fencing – Sweet meat – Bags – Basket – Firewood – Wrapper – Rope – Vinegar – Edible fruits – Hat	– Mat – Rain coat – Alcohol – Address poverty – Medicinal use (Vermicide, analgesic) – Fishing – Wine & toddy – Transport fuel (potential use)
Keora	– Cot – Fuel wood – Showcase – Chair – Fencing	– Table – Roofing – Ceiling – House column
Hargoza	– Fuel wood	
Possur	– Cot – Almirah – Showcase – Ceiling	– House column – Chair – Table – Roofing
Baen	– Fencing – Box – Roofing – Cot – Fuel wood	

Source Prepared by authors based on Unnayan Onneshan (2010)

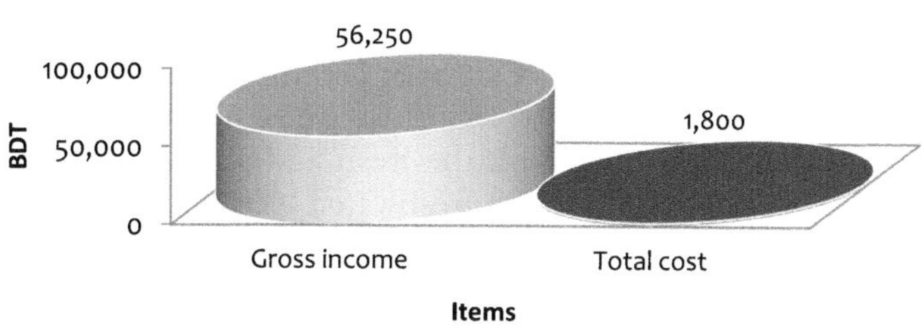

Fig. 9.9 Average yearly total cost and income in community-based mangrove silvi-culture (*Source* Authors' calculation based on Unnayan Onneshan [2010])

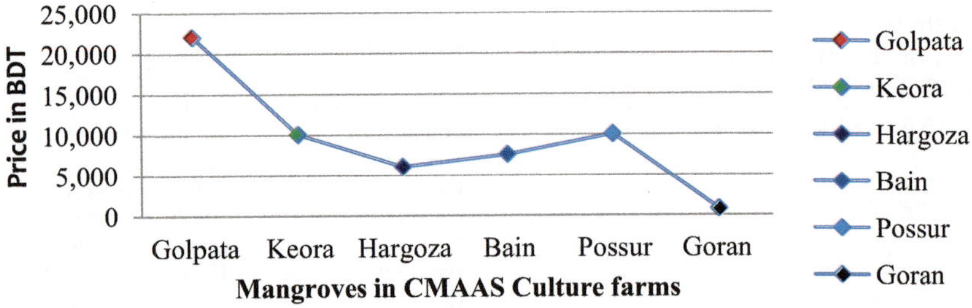

Fig. 9.10 Species-wise average yearly gross income from mangrove cultivation (per *bigha*) (*Note* A *bigha* is a unit of land measurement that equals to 1,600 yd^2 [0.1338 hectares or 0.3306 acres] and is often interpreted as being 1/3 acre [precisely 40/121 acres]. In metric units, one *bigha* is hence 1,333 m^2. *Source* Authors' calculation based on Unnayan Onneshan [2010])

greater contributor to the gross income of community mangrove cultivation is Golpata.

Cultivation Method of Aquatic Species Under CMAAS

Aqua-faunal species' cultivation is one of the parts of CMAAS culture, which consists of two processes. In general, the silvi fishery combined cultivation farms depend on one another in proportion to silvi and fishery cultivation coverage. The extent of variation depends from country to country and also from one farming model to another. Some farms are silviculture dominated while others are aquaculture focused. In Bangladesh, the CMAAS culture farms are aquaculture prevalent. The aquatic cultivation consists of both the fresh and saline water species of molluscs, crustaceans and fish. The freshwater species of molluscs, crustaceans and fish are cultivated in freshwater farms, whereas the saline water species of fish, molluscs and crustaceans are cultivated in the brackish water farms.

Aquatic Species Cultivated in Farms

The CMAAS culture farms generally experience the cultivation of a few species of shrimps and fish. Bagda and Golda are shrimp species, whereas

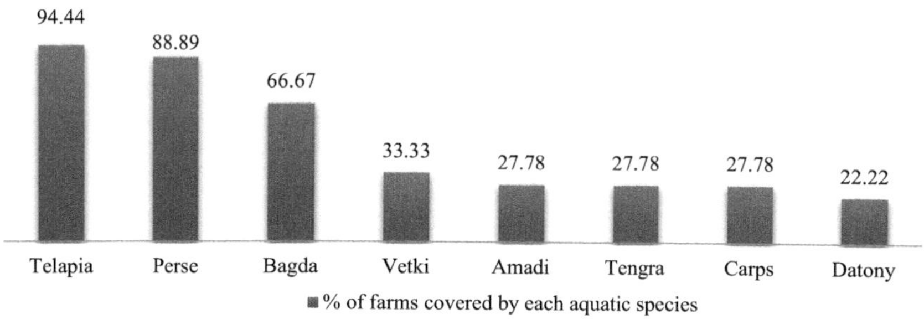

Fig. 9.11 Percent of farms cultivating different types of aquatic species (*Source* Authors' calculation based on Unnayan Onneshan [2010])

Telapia (*Tilapia nilotica*), Perse, Vetki (*Lates calcarifer*), Tengra (*Mystus tengara*), Baila (*Awaous guamensis*), Perse, Amadi, Carps and Datony are fish species.

Figure 9.11 shows that there are three types of fish like Telapia, Perse and Bagda which are commonly cultured in the farms. Among them, Telapia is the most cultivated fish in the CMAAS culture farms. It is cultivated at about 94.44% of the total farms. Besides, Perse and Bagda, there are two other widely farmed fish species that cover about 88.89 and 66.67% of the farms, respectively. The fish species which are least cultivated in the farms are Vetki, Amadi, Tengra, Carps and Datony which almost equally exist in the farms.

Fry Collection

The farming is entirely based on fish and shrimp fry being released into the farms regularly. They collect the fry in three ways, that is, self-collection from the rivers, buying fry collected from rivers and growing them in hatcheries. About 22% of the farmers, particularly the marginal ones, catch the wild-fry directly from the coastal rivers, while another 18% buy it from nearby hatcheries. The rest 60% farmers buy the fry from local sellers who collect them from the rivers. Catching fish fry and shrimp larvae in the coastal rivers are common sights in the south-west region of Bangladesh.

Table 9.3 The farming seasonality of the fish species cultivation

Name of fish	*Period of the release of fries*	*Fish harvesting period*	*Time needed for harvest*
Fish species of brackish water farm			
Telapia	December	April–May	3 months
Parse	February	May	2.5 months
Bagda	February	June	4 months
Vetki	February	May	3 months
Amadi	January	May	4 months
Fish species in fresh water farms			
Tengra	February	May	3 months
Carp	December	August	8 months
Datony	February	June	4 months

Source Prepared by authors based on Unnayan Onneshan (2010)

Seasonality of Farming

The seasonality of fish and shrimp growing at the CMAAS culture farms varies per species. Fries that release periods of all fish and shrimp species in the farm are primarily limited to the months of December to February each year. Similarly, the harvesting season is nearly same with minimal variance and is mostly confined to May and June. This is due to the fact that practically all fish and shrimp species spawn at the same time. Furthermore, from late May onwards, tropical cyclones with tidal surges and tidal floods are highly likely to occur. If such a natural disaster occurs, the farms' fish and shrimp are washed away, incurring a substantial loss to the producers. This is why, in most areas, fish and shrimp are planted before May to reduce losses due to natural catastrophes, as shown in Table 9.3.

Care and Maintenance

When compared to other sorts of fish farming methods, aquafarming in CMAAS culture does not necessitate as much care and maintenance. This method of farming has very low maintenance costs. Farm water does not have to be exchanged at regular intervals. Water can be changed whenever required. During the whole growth period, shrimp and fish are normally fed natural feed. Also, there are shrimp and fish feeds for purchase. Rice bran is the best homemade fish feed. The majority of farmers rely on natural feeding methods and rarely utilise chemical fertilisers on their farms. However, the farms must be monitored on a regular basis. This is to prevent intruders from catching fish and shrimp or cutting off the mangroves. The farm's dykes need to be heightened from time to time as they erode. In general, there are not many efforts made to fence farms. Nevertheless, isolated farms must effectively be fenced in areas where fish and shrimp are at risk of being stolen. Viruses can infect fish and shrimp sometimes, necessitating the use of medications and pesticides to heal them. Care is provided till they reach the appropriate size for commercialisation. The average annual cost of aquafarming care and maintenance is roughly BDT 5000 per *bigha*.

Figure 9.12 depicts the aquatic species' overall production cycle at the CMAAS farm. It shows how aquafarming procedure is classified into three types: oyster, crab and fish and shrimp. Among which oysters and crabs have a natural intake of 100% which then can be sent for care and maintenance. On the other hand, fish and shrimp cultivation takes on fry collection. This can be done in three ways: buying fry grown in a hatchery, buying the fry caught from rivers and self-collection from rivers. All three will help to release fry into the farm which can then be sent off for care and maintenance. Then, it can be used for harvest and finally for processing and distribution.

Harvesting

Almost all fish and shrimp require 3–4 months of growth following the discharge of their fry before they are marketable. In reality, harvesting and releasing occur in a cyclical fashion. Following the first stocking of fry, partial harvesting of marketable shrimp begins. Perse and Carps are the fish species with the shortest and longest periods.

Cost and Return on Investment

Aquafarming generates an annual average income of around BDT 1,830,000 per *bigha*. The entire cost, on the other hand, is just BDT 14,750, with a net profit of around BDT 1,73,250 (Fig. 9.13). As a result, the cost–benefit ratio is 1:12.

Furthermore, the annual average income of all aquatic species gathered in CMAAS culture farms is not the same. Rather, it varies greatly from one species to another. According to Fig. 9.14, the biggest contributions to CMAAS's gross average yearly revenue are Bagda, Golda, Perse and Vetki,

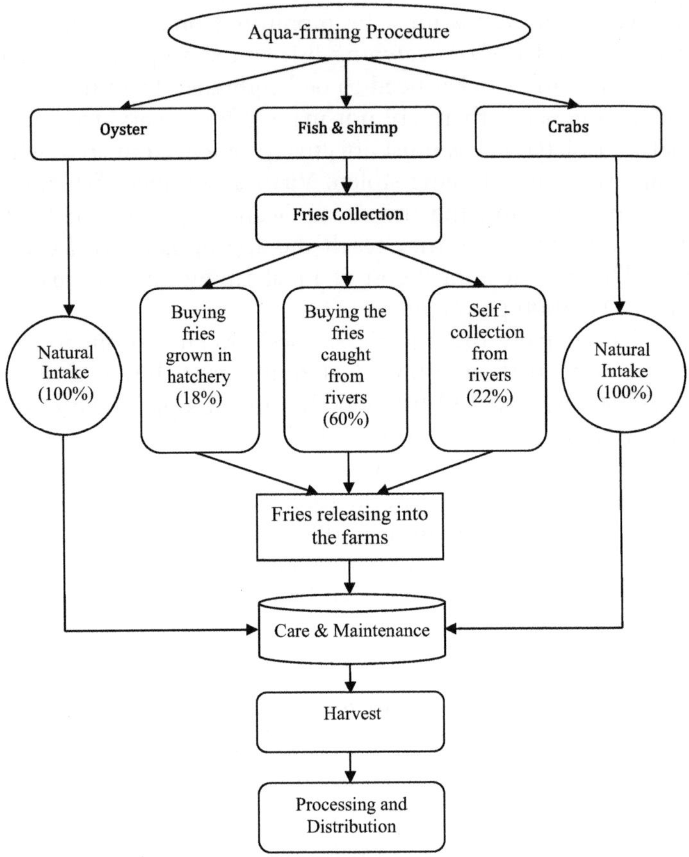

Fig. 9.12 Production cycle of aquatic species in CMAAS farm (*Source* Adapted from data reservoir of *Unnayan Onneshan*)

although Bagda cultivation generates the greatest yearly income, followed by Vetki.

The summary of economic returns of both flora and fauna species under the CMAAS culture is presented in Table 9.4.

Several variables have contributed to the formation of this profitable scenario. It does not need much care and maintenance because it relies heavily on natural feeding of fish and shrimp and the minimum use of fertilizer. The most essential consideration, however, is the cost of

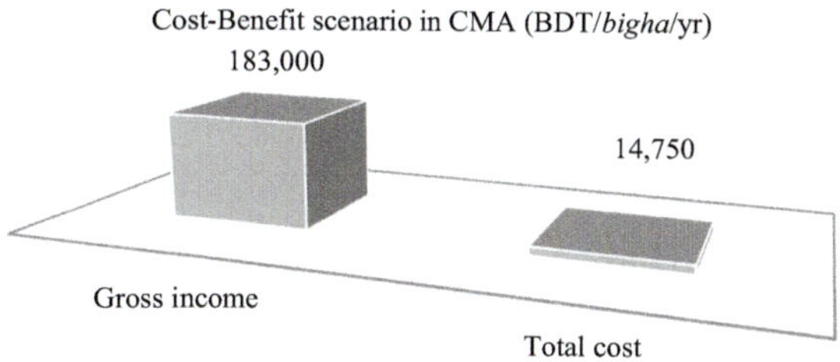

Fig. 9.13 Average yearly total cost and income in community-based mangrove aquafarming (*Source* Authors' calculation based on Unnayan Onneshan [2010])

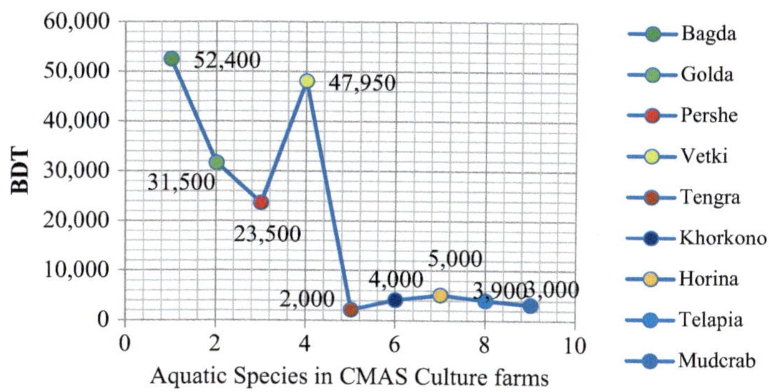

Fig. 9.14 Species-wise average yearly gross income from aquafarming (per *bigha*) (*Source* Authors' calculation based on Unnayan Onneshan [2010])

Table 9.4 Economic returns of CMAAS culture

CMAAS		
Economic Returns (Benefits > Cost)	**Mangrove cultivation (flora)** Total income (per *bigha*/per year): BDT 56,250 Total cost (per *bigha*/per year): BDT 1,800 Net benefit: BDT 54,450 Cost Benefit Ratio: 1:32	**Mangrove aquafarming (fauna)** Total income (per *bigha*/per year): BDT 183,000 Total cost (per *bigha*/per year): BDT 14,750 Net benefit: BDT 173,250 Cost-Benefit Ratio: 1:12

Source Prepared by authors based on Unnayan Onneshan (2010)

Table 9.5 Value of BCA measures of CMAAS and CS culture

Measures of CBA	*CMAAS Culture (BDT/*bigha*/yr)*	*CS Culture (BDT/*bigha*/yr)*
Present Value of Costs (PVC)	16,550	8,860
Present Value of Benefits (PVB)	217,500	177,272.72
Net Present Value (NPV)	202,454.54	169,218.18
Net benefit (NB)	200,950	168,412.72
Benefit-Cost Ratio (BCR)	13	20

Source Prepared by authors based on Unnayan Onneshan (2010)

labour. CMAAS culture farms must be patrolled at all hours of the day and night. In general, self-labor is used in this case, and labor costs are not assessed in monetary terms, which significantly reduces the total cost. Besides, both parts of CMAAS culture—community mangrove aqua-firming and community mangrove cultivation—share much of the care and maintenance costs.

Comparative Analysis of CMAAS Culture and Commercial Shrimp (CS) Culture

A comparative analysis of the two different forms of culture is summarised below. Table 9.5 depicts the comparison in economic terms.[1] CMAAS culture appears to be more profitable than CS culture in terms of Net Present Value (NPV) and Net Benefit (NB). When evaluating the Benefit Cost Ratio (BCR), however, the situation is considerably different. According to the BCR scenario, the cost effectiveness of CS culture is somewhat greater. Shrimp cultivation is, therefore, unquestionably profitable. But, beneficiaries in this case are a small group of people who have harmed the livelihoods of landless and marginal farmers.

Moreover, the ecological comparison (Table 9.6) revealed that the CS culture is extremely harmful to the environment. On the contrary, the CMAAS culture has little or no negative influence on the environment. It preserves the lands and soil from erosion, improves the use of fallow lands, protects the environment from pollution, aids in the conservation

[1] The Cost-Benefit Analysis (CBA) approach was used to compare the economic returns in this case.

Table 9.6 Ecological comparison between CMAAS and CS culture

Criteria	*CS culture*	*CMAAS Culture*
Salinity	Increases salinity in soil (in farmland and adjacent lands)	No use of saline water; no salinity intrusion
Use of lands	Used ponds exhaust usefulness within three to six years of construction. So, the destruction of mangroves occurs to make room for more ponds	Homestead adjacent fallow lands are used and no conversion of forest lands into cultivation lands
Use of chemical fertiliser, pesticides, insecticides	Chemical fertilisers, insecticides, etc. are used, causing pollution	Chemical fertilisers or insecticides are not used, natural feeding and therefore, no pollution
Impact on agricultural productivity	Limits crop production in agricultural land (by increasing salinity of lands) and conversion of agricultural lands to shrimp farming ponds reduces land availability	The agricultural productivity is unaffected
Impacts on the Sundarbans (in particular)	Eradication of natural mangrove vegetation, and pollution of aquatic resources (negative)	Eases and reduces the increasing anthropogenic pressures, making an alternative source of livelihoods for the local people who are dependent on the Sundarbans
Adaptation to climate change	Increases the vulnerability to climate change	An innovative adaptation method to climate change for the vulnerable

Source Prepared by authors

of biodiversity resources and most significantly provides alternative and sustainable livelihood options.

The following section depicts a case study about the pioneer of the CMAAS Culture. It talks about a farmer named Khoybor Sardar and how he has executed the entire plan from scratch. It talks about the resources he used, the amount of land, money, time and effort that have been put into bringing this culture and practice to life.

Fig. 9.15 The Pioneer of CMAAS Culture—Khaibar Sardar in his farm (*Photo* ©Loban Rahman, Unnayan Onneshan)

Box 9.1 Case study of the CMAAS culture pioneer

Khoybor Sardar, 60, is a marginal farmer cum traditional collector (Bawali) of resources from the Sundarbans Mangrove Forest in the village of Nanksha, union of Amadi, upazila of Koyra and the district of Khulna. He has four sons and four daughters. He lives in an extended family with his wife and sons together, with their spouses and children.

The enthusiastic farmer pioneered the integrated cultivation of some mangrove species, both floral and aquatic, like Golpata, Keora, Goran, shrimp (Bagda) and some types of finfish (Paissha, Vetki, Bangal and the like) for CMAAS culture. Being self-motivated and using traditional knowledge and innovative ideas, he started practicing CMAAS culture about 10 years ago, as an experiment and eventually the initiative became fruitful. Now, he briefly narrates his success story here.

Mr Sardar had a small plot of lowland with submergence almost all the year round amounting to 12 decimals next to his homestead, which was hardly used for any productive purpose. He collected the Golpata seeds floating on the river water and grew saplings from them in a home-made seedbed and planted them on his farmland with knee-deep water. Then, he cultivated shrimp and finfish mentioned thereof in that marshy land. After only five months of cultivation, he started getting the harvest. Now, his yearly earnings from Golpata and fish is about BDT 6,000–8,000 and 12,000–15,000 respectively with a cost of about BDT 2,000–3,000 only for fish culture and minimum care and maintenance. In addition, a few Keora trees have added a new look and economic window to the farm. Interestingly, one of the trees has borne a large beehive which quarterly offers about 3–5 kg honey fetching a market value of about BDT 1,000–1,500. It is a successful innovation of his life, as he claims.

On 25 May 2009, the Cyclonic Storm Aila harshly hit the coastal part of Bangladesh, widely breaching the coastal embankment. Consequently, almost the whole of Koyra Upazila was inundated by the saline water of the sea. The agricultural lands lost growing traditional crop varieties.

Mr Sardar, who started his farm after the cyclone Aila, argued that in today's context, CMAAS culture may be one of the best adaptation measures here in the Aila impact coastal zone if it can be effectively disseminated after studying its further feasibility (Fig. 9.15).

Conclusions

This chapter emphasised CMAAS culture, starting from its emergence to how it was executed and brought to life, in addition to all the benefits it has to offer. Numerous cultivation methods were accounted and explained thoroughly. The main purpose of this chapter is to register the CMAAS culture and the various sectors and sub-sectors that are related to it. Each type of cultivation, whether that of mangrove or the aquatic species were shown with details that provided an overall understanding. In addition, natural and anthropogenic drivers, impacts and responses to climate change and salinity intrusion in the coastal area and their linkages to the emergence of CMAAS culture are shown to be interlinked with one another. The different cultivation methods under CMAAS culture have also been depicted along with distinctive ways of resources that are being used. This makes it easier to compare the two to sort out the impacts and outcomes of each method. In addition, cultivation methods which consist of economic costs, average yearly total cost and income, etc. have also been given.

The CMAAS culture is economically profitable, environmentally supportive and socially acceptable than the Commercial Shrimp (CS) culture in the coastal region of the country. This has been proven after a comparative analysis of CMAAS culture and Commercial Shrimp (CS) cultivation. Indeed, it is found to be the best alternative to Commercial Shrimp (CS) cultivation, as it is more profitable and has a lesser negative impact on the local environment and biodiversity. CMAAS culture deserves to be suggested as one of the finest coastal adaptation techniques, taking into account the physiographic and socioeconomic characteristics, as well as the vulnerabilities and catastrophe statistics of the coastal region. As a result, embracing CMAAS culture will encourage improved resource use while having a minimal negative influence on the forest's biodiversity and environment. It has been deemed an acceptable and practicable approach.

Overall, the CMAAS culture is a unique adaptation strategy in the face of climate change in the coastal region and the local people have developed it, providing a strong ownership and a scope for expansion. For this reason, further rigorous study is required to explore its feasibility through extended field experiment and research.

Acknowledgement This chapter has mainly been prepared based on the research findings of a study conducted as a project of the Unnayan Onneshan under the supervision of Dr. Rashed Al Mahmud Titumir, and the research team included Mohammad Abdul Baten and Md. Humayun Kabir who previously worked as researchers at the Unnayan Onneshan

References

Baten, M. A, & Kumar, U. (2010, May 3–5). *Responses to the changes in the Sundarbans*. Paper presented at international conference on biological and cultural diversity, UNCBD.

Dasgupta, S., Sobhan, I., & Wheeler, D. (2017). The impact of climate change and aquatic salinization on mangrove species in the Bangladesh Sundarbans. *Ambio*. https://doi.org/10.1007/s13280-017-0911-0

Falk, G. (2015). Land use change in coastal regions of Bangladesh: A critical discussion of the impacts on delta-morphodynamics, ecology and society. *The German Journal on Contemporary Asia, 134*, 34–73.

Islam, M. S. N., & Gnauck, A. (2009). Threats to the Sundarbans mangrove wetland ecosystems from transboundary water allocation in the Ganges basin: A preliminary problem analysis. *International Journal of Ecological Economics and Statistics, 13*(9), 64–78.

IUCN. (2014). *Bangladesh Sundarban delta vision 2050: A first step in its formulation*. Bangladesh Country Office.

Kabir, D. M. H., & Hossain, J. (2008). *Resuscitating the Sundarbans: Customary use of biodiversity and traditional cultural practices in Bangladesh*. Unnyan Onneshan, BELA, Forest Peoples Programme and Nijera Kori, Dhaka.

Parvin, G., Ali, Md. H., Fujita, K., Abedin, Md. A., Habiba, U., & Shaw, R. (2016). *Land use change in Southwestern Coastal Bangladesh: Consequence to food and water supply*. https://doi.org/10.1007/978-4-431-56442-3_20

Parvin, G. A., Takashi, F., & Shaw, R. (2009). Coastal hazards and community-coping methods in Bangladesh. *Journal of Coastal Conservation, 12*(4), 181–193.

Rezaie, A. M. (2015). *Quasi-real time prediction of storm surge inundation for the coastal region of Bangladesh*. Master's thesis, Institute of Water and Flood Management, Bangladesh University of Engineering and Technology, Dhaka, Bangladesh. http://lib.buet.ac.bd:8080/xmlui/handle/123456789/4062. Last accessed on 25 March 2021.

Sarker, S. K., Reeve, R., Thompson, J., Paul, N. K. & Matthiopoulos, J. (2016). Are we failing to protect threatened mangroves in the Sundarbans world heritage ecosystem? *Scientific Reports*, *6*(21234). https://doi.org/10.1038/srep21234

Titumir, R. A. M. (2011). *Property rights, CSU and development*. Paper presented at the Expert Committee meeting on CBD article 8(j), CBD Secretariat.

Titumir, R. A. M. (2015, November 6). *IPLCs contribution to Aichi biodiversity target 10: A case illustrated through TRUs of Sundarbans in Bangladesh*. Paper presented at side event on indigenous peoples and local communities' contributions to the implementation of the strategic plan for biodiversity 2011–2020.

Titumir, R. A. M., & Afrin, T. (2017). Complementarities of human-nature well-beings: A case illustrated through traditional forest resource users of Sundarbans in Bangladesh. *Satoyama Initiative Thematic Review (SITR)—Volume 3*, United Nations University Institute for the Advanced Study of Sustainability (UNU-IAS): Tokyo, Japan.

Unnayan Onneshan (UO). (2010). *Community based mangrove agro aqua silvi (CMAAS) culture: Promoting as a community adaptation tool and an alternative to commercial shrimp culture*. Research report. Unnayan Onneshan.

World Bank. (2016). *Impact of climate change and aquatic salinization on mangrove species and poor communities in the Bangladesh Sundarbans* (World Bank Policy Research Paper 7736).

CHAPTER 10

Traditional Resource Users and Claiming of Rights

Rashed Al Mahmud Titumir, Mostafa Walid Pasha, and Md. Shah Paran

Introduction

This chapter thoroughly explores the successful activities and the achievements of three cooperatives functional in the Koyra region, located in the Khulna district of Bangladesh. An independent multidisciplinary research think-tank, the Unnayan Onneshan (UO), has been working in this region since 2010. In 2011, after rigorous efforts, the organisation was able to set up three development cooperatives (Koyra, Harinagar and Munda Indigenous) comprising three different communities of local and

R. A. M. Titumir (✉)
Department of Development Studies, University of Dhaka, Dhaka, Bangladesh
e-mail: rt@du.ac.bd; rtitumir@unnayan.org

R. A. M. Titumir · M. W. Pasha · Md. S. Paran
Unnayan Onneshan, Dhaka, Bangladesh

R. A. M. Titumir (ed.), *Sundarbans and its Ecosystem Services*, Sustainable Development Goals Series,
https://doi.org/10.1007/978-981-19-3000-3_10

indigenous people. Currently, around 350 households are enlisted as members of these cooperatives.

Forming a cooperative is an old concept, in which organisations are set up keeping in mind certain goals. Here, the members work in cooperation bringing together different perspectives to obtain a common goal. Any organisation has certain rights which they need to secure to further mobilise their organisation. Rights can range from being environmental, social, economic, gender-related and others, based on their activities and goals. For example, cooperatives may need to claim their property rights to attain their goals. Every individual should know when to press for their rights for which putting pressure and negotiating with concerned authorities becomes significant. Cooperative organisations garner power by mobilising individuals to come together for a greater good and work for a manifested goal and make it functional. Such movements bring change. Precisely, securing the rights of an individual, especially by organising them and forming cooperatives is crucial for changing their condition.

Cooperatives are formed to serve the interest of different sectors such as workers, employees, farmers, forest people and others. The success and failure of cooperatives depend on certain factors, which determine if the organisation will be able to serve a fruitful outcome for the members of the cooperative. If cooperatives cannot perform for the betterment of the common people for whom these programmes are introduced, they become valueless. Many cooperatives fail due to conflict of interest of their members, lack of proper knowledge, profiteering, individual promotion, class discrimination and duality in the role as both member and owner, structural change of the concomitant environment and neoclassical attributes.

These factors set them adrift from the welfare of the general public, towards individual profiteering. A cooperative is supposed to function democratically and maintain equality to provide multi-dimensional facilities to its members. It can work as a bank where the members deposit their money and invest them in a business in the future. The generated profit is not collected initially. Instead, it is reinvested for future development. There are some established values of cooperatives such as self-help, self-responsibility, democracy, equality, equity and solidarity. Cooperatives have certain fundamentals and principles such as voluntary and open membership, democratic member control, member's economic participation, autonomy and independence, education, training and information, concern for community and others.[1]

The forest people cooperatives in Koyra region were set up to boost their capability in terms of knowledge, life skill training, entrepreneurship, awareness building and creating negotiating capacity as well as an ambience of opportunities for improving their socio-economic condition. Remarkably, the TRUs have been conserving the natural biodiversity through their sustainable customary traditional practices. The purpose of protecting the ecology of the forest is automatically included in these cooperative's function.

Activities and Achievements

This section provides successful case studies from the field. These case studies have been obtained by the Koyra, Harinagar and Munda forest people multipurpose development cooperatives of Koyra upazilla, in Khulna, Bangladesh. These studies illustrate that cooperatives can perform effectively if they are united, have proper information, knowledge and training; and are determined to achieve their goal. Since commencement, these cooperatives have excelled through proper knowledge, skills and uniformity, in terms of raising awareness and taking decisions for their universal welfare.

The gradual boost in the movement has improved their negotiating power and pressure imposing capability (Fig. 10.1). People have claimed their rights by the congregated efforts of cooperatives by mobilising the masses for movement during the period with well-organised activities.

Mobilisation Effect

Cooperatives have been identified as a 'special kind of business' for their contested existence in the capitalist environment since inception (Kramper, 2012). Cooperatives may succeed or fail depending on their contemporary ambience, policies and regulation patterns. If the cooperatives are influenced by capitalist behaviour, the key objective of its formation may get diluted over time. Cooperatives fail in two ways,

[1] Available at https://www.ica.coop/en/cooperatives/cooperative-identity?_ga=2.190013346.849363065.1585542157-1500733193.1585542157.

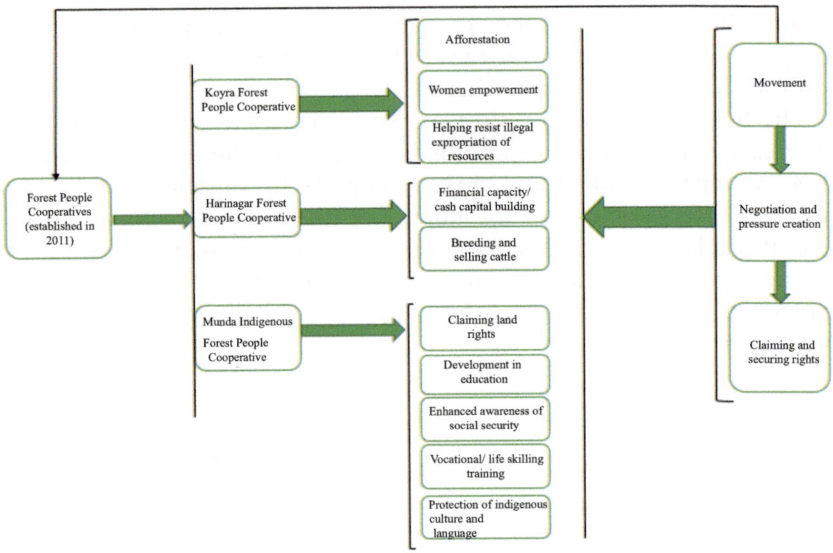

Fig. 10.1 Diagram of claiming rights through cooperative functioning (*Source* Prepared by authors)

one by getting bankrupt and the other by transforming into investor-owned businesses (Kramper, 2012). There have been debatable thoughts on the distribution and success of cooperatives. A producer cooperative flourishes rarely and even if it succeeds, it ceases to be a producer cooperative (Oppenheimer, 1896 cited in Kramper, 2012). From a Marxist perspective, it can be said that isolated worker cooperatives fail to bring changes, to the wider forces and relations of production that have developed under capitalism (Oppenheimer, 1896 cited in Kramper, 2012). In short, a cooperative itself has to survive in a competitive market economy to maximise profit like other profit-making organisations.

In a neoclassical environment, worker cooperatives may not be able to survive due to their capitalist nature and rent-seeking behaviour in the form of increasing net income per worker. This creates a negatively sloped short-term supply curve and in an incorrect response to changing market conditions (Ward, 1958). There might be problems originating inside the cooperative due to the conflicting interests of the members, their diverse and inadequate tradability and the dual role of the members as owners and users. In new institutional terms, this problem is called the 'horizon problem' due to the limited patronage horizon of the members (Fulton, 2001). Very often members regard cooperatives as a form of an investment institution since they get their returns against the investments in a short-term succession. Thus, they deviate from long-term planning to short-term objectives in respect of a return on investment (Fulton, 2001) (Fig. 10.2).

In such a situation, the formation and functions of the three cooperatives have been quite exhausting and hazardous. There were many difficulties that are still prevailing in the society of Koyra, Harinagar and Munda forest people. They face problems while collecting resources from formal institutions such as the local government and the forest department. Influence and muscle power wielded by some powerful people, who are mainly rent-seeking and expropriating in nature, create further trouble for the common people. Despite these troubles, the forest people responded to the UO when they received the proposal of making a cooperative for their own betterment which would be sustainable for them.

Fig. 10.2 Banner of third Sundarbans biannual forest peoples convention (*Photo* Unnayan Onneshan)

Since its inception, there has been continuous progress with positive outcomes in various arenas. For example, Koyra forest people cooperative led to massive afforestation drive which caused a vivid impact by making their surroundings healthier than they were earlier. They protected their houses and roads by planning on river isles. Soil erosion frequency was stalled. Harinagar forest people cooperative made remarkable success in improving their financial capabilities by allowing deposits and lending of money for small and medium level investments. Munda forest people cooperative attained tremendous community knowledge, vocational training and honed their skills which are important for community development. As a matter of fact, these cooperatives became aware of their rights and needs. They have learned how to compel the formal authority to provide facilities that are their rights and needs as a citizen. They have evolved and developed their negotiating power and gained knowledge regarding which door to knock and where to go to negotiate their issues. The ability to instil pressure over authorities to claim their fundamental rights have been obtained through the unity of the cooperative members, since its establishment in 2011.

Land Struggle

Cooperatives have to face difficulties dealing with property rights very often, especially those in the Koyra region. The people in these regions lost their land and had to struggle to claim it back (Titumir & Paran, 2022a). Initially, they failed but they stood up and fought back. Previously, they did not know how to protest against injustice that was burdening their lives. In fact, before the formation of the cooperatives, they lacked the courage to proceed and claim back their lost land. Gradually, members were empowered through continuous knowledge development sessions, awareness-raising programmes, educating illiterate people and demonstrating the way out.

Munda indigenous forest people multipurpose development cooperative struggled with the land crisis. This was due to the powerful people who had snatched away their inherited land with the help of some notorious people within the local government. Recently, they have been able to regain and restore some of their lost land which amounts to 42 bighas (29 acres), from the powerful encroachers. The reclaimed land is located between 22° 18′0″ north latitude and 89° 18′10″ east longitude (Fig. 10.3).

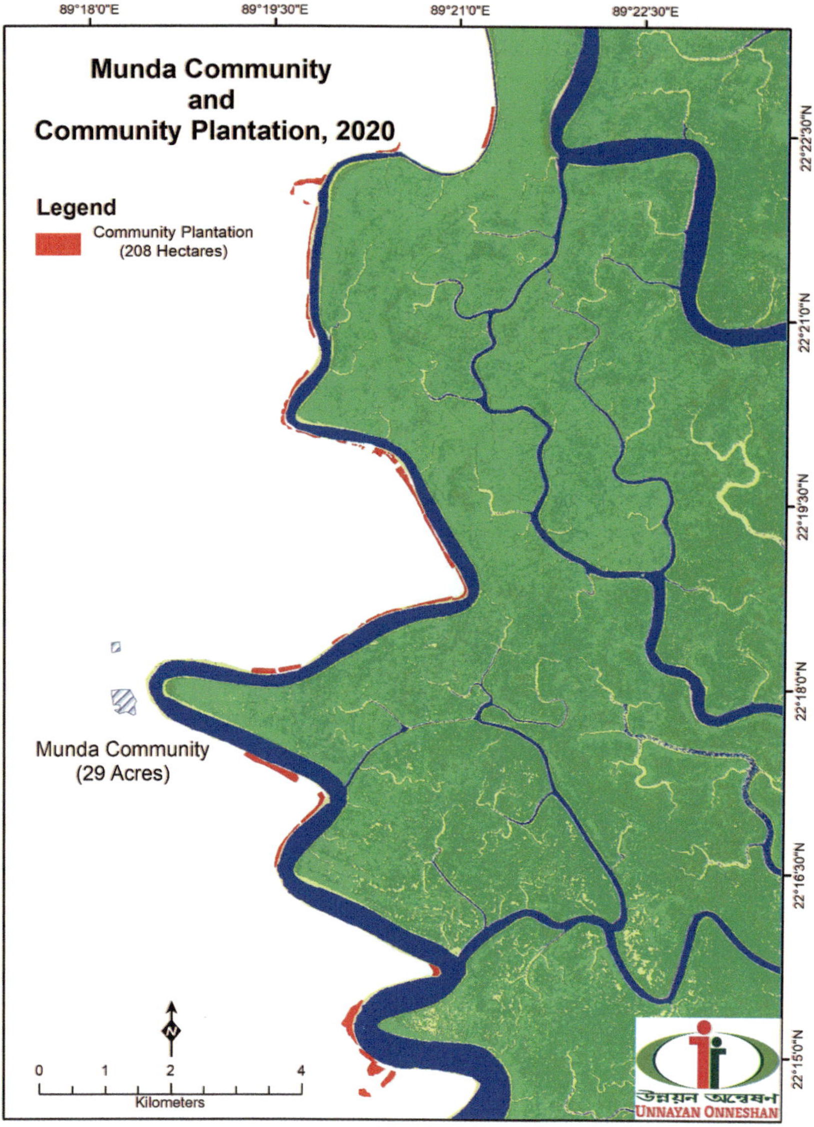

Fig. 10.3 Reclaimed land by Munda Indigenous people and community plantation in Koyra region (*Source* Unnayan Onneshan [2020])

The road towards regaining their land was not easy. They had to fight, get beaten up and face many threats from powerful cartels. The journey was risky and troublesome but it did not stop them from pursuing their goal. They got their legally owned land through a lot of effort. At first, they were able to regain 24 bighas of land after facing several conflicts which even resulted in a head-on collision with the ones colluding with power. Later, they achieved 5 more bighas after overcoming several threats, court cases and harassment by the local government and politicians. At a point in time despite being backed by a court order they failed to gain access to their land and faced continuous threats. Joydeb Munda, president of Munda forest people cooperative said,

> We knew the journey was not easy. They tried to squeeze us as much as they could but we were adamant to attain our legal rights.

Some personnel were directly responsible for the deprivation of the indigenous Munda forest people.

Despite threats of being sent to jail, the determination of the Munda cooperative members led them to directly negotiate with the local government and police. Several lawsuits are still pending in the court but they are determined to regain their land and distribute it properly among the cooperative members. The major reasons behind their land movement were their well-polished ideas regarding the goals, proper knowledge and training and united efforts to induce congregated pressure on the local government.

Negotiation with the Local Government

The success of a movement is based on the ability of the participants of the movement. The more they can create pressure on the authority or holding power to ratify their demand, the more likely they are to succeed. At the end of the day, the success of any social movement depends on the efficacy of its leaders. Leaders who are committed to power sharing and leading from behind, generally are more capable and change-making. Indeed, a 'leadership' movement successfully nurtures and empowers the energy of many and transforms that energy into a common cause (Crutchfield, 2018). Movements should be of common interest which go in favour of the common people, not a few profit-seeking vested interest groups. When a movement gets captivated by the neo-liberal spectrum and profit becomes the optimum cause for a few of them, the result turns into negativity. Failure of movements in Thailand and South Korea implies the negative aspects of neo-liberal strategies that are driven by the market economy and also the lack of capability to internationalise their efforts to bring social change to some extent (Glassman et al., 2008). A movement should focus on universality that can enhance the capacity of every individual to create pressure on the authority to reclaim their needs and rights.

Munda indigenous forest people cooperatives made a signature achievement of having a negotiating capability. Negotiating capacity has paved the way for obtaining success. Several issues like conserving the resources of the mangrove forest, providing vocational training to the members of their communities, ensuring allowance for elderly people, mitigating domestic violence, afforestation, conserving and fostering the distinctive culture, language and tradition of the Munda community. In addition to the provision of public goods, distributing seeds among poor farmers, creating small entrepreneurship, crab culture, alternative conservation methods of the forest and many others have got a focus for further activities. In terms of these issues, Munda people have expanded the capacity of negotiating with the forest department, local politicians and police force. They have identified their prime needs and problems which hinder their progress towards sustainable development. After identifying their needs, experienced and well-educated Munda community members have assembled other members to make them understand their problems and crisis. Thus, unity grew among the Munda forest people cooperative which allowed them to present their needs to the incumbent administration. They stumbled, got threatened and were wounded both physically and financially. Yet, these phenomena taught them how to proceed towards achieving their goals. It is interesting to hear a Munda Indigenous woman say,

> We have learned how to proceed with our demands. We know where we should go and how we have to make a negotiation.

They even went to the district level authority to discuss, and send the message demonstrating their crisis and current condition. Knowledge and training are easier than before to make discussion and create pressure.

Conservation Practices

Munda indigenous forest people have been living around the forest region for many years. Like other communities, they have been involved in collecting resources from the forest. Thus, traditional indigenous people and the forest have been tied in a mutual relationship of interdependence and well-being (see Titumir et al., 2019; Titumir & Paran, 2022a, 2022b, 2022c). They can feel that their activities affect the ecosystem cycle of the forest environment. They have been appropriating resources from the forest by catching fish, collecting honey, crab and *golpata* for several years. They now realise that even if they collect resources from the mangrove forest in a sustainable and environment-friendly manner, due to the intrusion of outsiders and powerful group of people along with local government, expropriation will occur. In this circumstance, they have decided to make alternative conservation by decreasing their dependence on the forest to maintain their socio-economic livelihood.

They opted for crab culture and fisheries on their lands which help them earn profits regularly and lessens the pressure on the forest. They use traditional methods while cultivating crabs and other aquatic resources. Moreover, they are now getting vocational training as well as training on proper farming which lets them conserve the soil and water of the forest region. Joydev Munda, the current President of the Munda Indigenous Forest People Cooperative, opines,

> Pressure on the forest should be minimised. We have to protect it. This feeling made us divert into other activities that would help contribute to conserving the resources of the forest.

They conduct afforestation programmes regularly which are making both the forest and people beneficiary by protecting the trees, soil, water and air. Munda people have been putting their focus on education, conserving their distinctive culture and ensuring self-reliance goes beyond random dependence on the forest. In terms of crab culture, they do not catch the spawns so that they can breed properly. Apart from crab culture, they refrain from using harmful chemicals in other farming activities.

Community Plantation in Koyra

On the map, community afforestation has been done on 42.10% of 494 hectares' land along the embankment of Shakhbaria river (Rai river) in Koyra upazila (Fig. 10.4). The area of this afforestation along the embankment is 206 hectares. The foresters have collected the seeds floating in the river from the trees of the Sundarbans and planted them in their own lands. In this way, a huge area has been covered with green trees. This community afforestation is, on the one hand, protecting embankments and on the other hand, it is also working as a defence against frequent natural calamities. Moreover, various aquatic species including fish and crabs come up in these lands in the tidal period. They use the trees, fish, crabs, etc. of this part of their land themselves. As a result, reliance on forests is reduced, albeit slightly. This forestry is also working as a shelter for various birds.

Since the forest people of these cooperatives have flourished, they have been dependent on its resources for many years. They earn their livelihoods by extracting resources from the forest. As discussed earlier, their traditional method of collecting resources contributes to the conservation of the ecosystem's biodiversity (see chapter 4 & 5). Understanding the mutual dependence of existence between humans and nature was enough to ensure sustainable resource collecting practices in this region. However, this was until there was aggression by outsiders who started expropriating resources for their individual interest without considering the benefit

Fig. 10.4 The scenic Shakbaria river surrounding the Sundarbans region in Koyra (*Photo* Unnayan Onneshan)

and well-being of nature. Their expropriating nature has jeopardised the ecological balance of the forest.

The TRUs who are also members of the cooperatives are inherently aware of the importance of the conservation of their mother forest. Especially, Koyra forest people cooperative has deemed alternative ways to lessen their dependence on the forest. Apart from protecting their houses and roads from river erosion, they have planted several trees on the isles of the riverbanks alongside their house premises. Khalil Dhali, the president of Koyra forest people cooperative, says.

> We want to save our forest. If forest exists, we will exist. Plantation of trees in our living areas will reduce pressure from the Sundarbans.

They collected seeds and planted trees which are mainly born in the forest. As a result, soil and river erosion declined, birds come and sing and they also inhale fresh air like they used to take several years ago.

Education and Health

All three cooperatives now feel the necessity of attaining education. They know that without acquiring proper education, they cannot move forward to secure their rights and needs. Among them, Munda forest people cooperative prospered more than the others in terms of education. Rising consciousness and campaigns by previously educated people have created an ambience for maintaining the light of education. Recently, it has been found that some of their members have also completed graduation degrees from public universities and college degrees from different areas. These educated people are working hard to spread the importance of education from the very primary level. Apart from educating their offspring, they are also concentrating on adult literacy schools. These are for those who could not attain education when required due to several socio-economic hindrances. Munda Indigenous cooperative president said,

> We have gradually felt the necessity of education. Now our cooperative is working hard to ensure education from the primary level to the tertiary one. Educational attainment will improve our socio-economic status.

The cooperative has made some specific plans. These include buying necessary equipment for education to assure proper nutrition while attending schools. Apart from educational attributes, they are now showing more concern about health. This is because the area where they live is quite remote and health facilities are not easily available due to the absence of adequate clinical provisions and good doctors. They are now negotiating with formal authority to make them assure maternal health, safe delivery of children and enhanced nutrition for everyone. Moreover, adolescent health has also got their concentration which is worth appreciating.

Financial Capability

Raising financial capability among the community members is another amazing factor that has been found while inspecting the current socio-economic condition of the three cooperatives. Harinagar forest people cooperative has been able to improve financial capability through community deposits and provision of borrowing money from the cooperatives to use this in small and medium level investments. Throughout this process, the cooperative has been working as a bank. This is where people can deposit money and receive money from the cooperative on easy conditions after a particular period. This mode of financial mobilisation has coordinated socio-economic development since small and medium level farmers and other workers can get capital to use in their respective production processes. Jamir Sardar, president of Harinagar forest people cooperative, opined.

> Our community people are getting self-reliant and more confident than before. They are now producing more crops, breeding more animals and doing many forms of business since capital has become more available through cooperative banking.

Social Security and Awareness

The latest concept of social security and sprouting awareness as an aptitude is visible among the cooperative members during the 3rd biannual forest people's convention. Among all, a rigorous and updated understanding of social security by the Munda forest people cooperative was prominent. As mentioned earlier, their educational advancement has created many opportunities which were exotic and unknown to them before. One of which is social security. They now feel the importance of the welfare of every member of the community and believe that every member should be provided with a basic income grant under the supervision of a cooperative. Apart from the demand for regular allowance for elderly people, they now express the importance of universal welfare through access to education, health and employment opportunities. The cooperative members believe,

> Social security will ameliorate our condition irrespective of socio-economic status. We need easy access to education and health. Otherwise, inequality will always exist.

In addition to this, members of the cooperatives have nurtured a practice of awareness against injustice, crime, deprivation, narcotics, child marriage and many other malpractices which disrupt the social harmony of their communities. Now they retain direct communication with the police force and have forced them to believe that their opinion matters.

Vocational Training

The reality that vocational and lifelong training should be prioritised is gaining importance in the mindset of the people. All the three cooperatives are now mentioning consistently that their vocational training opportunities should be increased. With the limited opportunities, several members have acquired computer skills, driving proficiency, alternative conservation methods and many others that they use in their everyday life. They are now planning to buy 'easy bikes' for creating more employment opportunities for their community members. They believe that these skills will improve their capabilities to earn more and thereby making them self-reliant. Yet, they have been facing shortcomings like scarcity of computers, absence of skilled trainers and cooperation of the government. The distance they have surpassed with these limited vocational skills has and would further increase the opportunities available. They believe that these will make them more successful in terms of getting self-dependent and solving several problems.

Women Empowerment

Keeping women in the dark is like fighting a war without weapons. The gradual realisation of utilising the capabilities of women is rising. Recently, cooperatives have given more space to their female members and have included them as their frontline warriors. Women now express their opinion more openly than before. Anju Ara, a confident and hardworking member of Koyra forest people cooperative said,

> I do not feel shy to speak loudly. As a human being, I feel the necessity of coming forward to speak for my rights, problems and the injustice and crime that I face and witness regularly.

They now represent their cooperatives in formal programmes such as conferences, seminars or conventions. They understand their problems thoroughly. For many years, women have been playing their role as active wage earners by catching fish from the rivers and harvesting in their crop fields. A fisherwoman, Jamila Khatun, feels that a more favourable environment should be created for women. She believes,

> We still face difficulties while going to the forest or catching fish from the rivers. Outsiders who use poisonous nets should be stopped and punished. Otherwise, our efforts to succeed as a cooperative will be nipped in the bud.

Protection of Indigenous Culture and Language

Among the three communities, Munda forest people are the only indigenous group of people who have a distinct culture and language. They have been evoking for a long time for formal cooperation from the local government to help them protect their own culture, values, norms and language. Previously, they faced a direct hostile attitude from personnel of district administration which seemed to be a threat to the existence of their cultural attributes. Joydev further adds,

> We went to the district administration office to talk about this matter but they hardly gave us the space to speak on this issue as if it had been valueless to them. We did not get daunted. We are now more united and confident and they also count on our determination.

Members of the Munda community believe that their own values, customs and language have been unrecognised and undervalued. Besides, due to the lack of practice of inherited language among the community members, their language is on the verge of extinction. In this respect, they have made some specific efforts such as increasing the use of their own language, adding Munda language in their primary education, formal recognition and initiative by administration to protect their culture, values, ritual and traditional practices and others.

Conclusions

Cooperatives work properly if the members have appropriate knowledge, training, innovative ideas and negotiating capabilities. Many cooperatives such as consumer cooperatives, producer cooperatives, hybrid cooperatives, credit cooperatives and others have failed due to problems arising from property rights, the dual role of members of the cooperatives, conflicting interest, class division, rent-seeking attitude, short term refund attainment expectation and many others. Some cooperatives could not flourish as they were unable to adapt to the changing socio-economic conditions. Very often cooperatives fail due to the profit maximising objectives of some members. The organisational inefficiency, structural change, market-centric approach and individual interest can be attributed to the negative outcome of many cooperatives.

Throughout the years since establishment, three cooperatives studied under this chapter have prospered in a satisfying manner, in diverse aspects by constantly learning, evolving and raising awareness of their necessities. They have come forward, discussed with each other, detected their vital problems and looked for possibilities, with which they can resolve their shortcomings. It is worth mentioning that the booming negotiating power and increase in pressure creating capability have contributed towards future development and a sustainable socio-economic status and provision of equity, social justice and equal opportunity.

Some people have regained their lost land, while some of them discovered alternative conservation strategies to protect their forest and enhance their earning opportunities with the help of the cooperatives. The importance of education, health, vocational training and social security has caught their highest attention. Another satisfying fact is the empowerment of women in their communities. Indigenous people have become more verbal and conscious in protecting their own culture and language.

Overall, cooperatives have learned a lot by uniting, learning their rights and needs by making a sense of the movement and negotiation with formal institutions.

A movement succeeds when the cooperatives work in favour of the needs and interests of the common people to assure universal and sustainable benefits. Addressing key problems, making discussions, allowing participation and vibrant opinions from the field level, ability to put pressure and negotiate with responsible institutions improves the possibility of succeeding by any movement operated by the cooperatives. These organisations should work transparently and should be represented by people from the grassroots level via proper knowledge and training. The activities and achievements of Koyra, Harinagar and Munda forest people cooperatives can be substantially considered to be a role model towards claiming and securing the rights of the cooperative members.

References

Crutchfield, L. R. (2018). *How change happens: Why Some social movements succeed while others don't*. Wiley.

Fulton, M. E. (2001). Traditional versus new generation cooperatives. In Christopher D. Merrett & Norman Walzer (Eds.), *A cooperative approach to local economic development* (pp. 11–24, here p. 19). Westport.

Glassman, J., Park, B. G., & Choi, Y. J. (2008). Failed internationalism and social movement decline: The cases of South Korea and Thailand. *Critical Asian Studies, 40*(3), 339–372.

Kramper, P. (2012). *Why cooperatives fail: Case studies from Europe, Japan, and the United States, 1950–2010* (pp. 126–149). Cambridge University Press.

Oppenheimer, F. (1896). *Die Siedlungsgenossenschaft*. Leipzig.

Titumir, R. A. M., & Paran, M. S. (2022a). Human-nature cooperation for well-being: Community understanding on one health approach in the COVID-19 era in the Sundarbans. In M. Nishi, S. M. Subramanian, & H. Gupta (Eds.), *Biodiversity-health-sustainability nexus in socio-ecological production landscapes and seascapes (SEPLS)*. Springer Nature.

Titumir, R. A. M., & Paran, M. S. (2022b). Ecosystem services and well-being in the Sundarbans of Bangladesh: A multiple evidence base trajectory. In R. Dasgupta, S. Hashimoto, & O. Saito (Eds.), *Assessing, mapping and modelling of mangrove ecosystem services*. Springer Sustainability Science Book Series.

Titumir, R. A. M., & Paran, M. S. (2022c). Human-nature relationship for post 2020 biodiversity framework: An exploration into traditional knowledge, customary sustainable use and community-based innovation in the

Sundarbans. In R. A. M Titumir (Ed.), *Fifty years of nation building: Political economy of Bangladesh's development*. Centennial Book Series Project of University of Dhaka. Department of Development Studies, University of Dhaka.

Titumir, R. A. M., Paran, M. S., & Pasha, M. W. (2019). The Sundarbans is our mind: An exploration into multiple values of nature in conversation with traditional resource users. In UNU-IAS & IGES (Eds.), *Understanding the multiple values associated with sustainable use in socio-ecological production landscapes and seascapes (SEPLS)* (Satoyama Initiative Thematic Review, Vol. 5, pp. 97–117). United Nations University Institute for the Advanced Study of Sustainability.

Unnayan Onneshan. (2020). *Biodiversity, climate change and traditional resource users in the Sundarbans: An exploration through Public Participation Geographic Information System* (PPGIS) (Unpublished Research Report). Unnayan Onneshan.

Ward, B. (1958). The firm in Illyria: Market Syndicalism. *American Economic Review, 48*, 566–589.

CHAPTER 11

Resilience and TRUs' Contribution Towards SDGs and Aichi Biodiversity Targets

Rashed Al Mahmud Titumir, *Tanjila Afrin*, *and Mohammad Saeed Islam*

Introduction

This chapter presents a model of sustainable conservation framework which can improve the resilience capacity of the Sundarbans' mangrove ecosystem. Resilience is used to define the tolerance capacity of a system to withstand shocks or disturbances and recover over time. The amalgamation of different ecosystems such as the forest, coastal and wetland make the mangrove forest home to numerous aquatic and terrestrial

R. A. M. Titumir (✉)
Department of Development Studies, University of Dhaka, Dhaka, Bangladesh
e-mail: rt@du.ac.bd; rtitumir@unnayan.org

Unnayan Onneshan, Dhaka, Bangladesh

T. Afrin · M. S. Islam
Department of Development Studies, Bangladesh University of Professionals, Dhaka, Bangladesh
e-mail: tanjila.afrin@bup.edu.bd

M. S. Islam
e-mail: saeed.islam@bup.edu.bd

R. A. M. Titumir (ed.), *Sundarbans and its Ecosystem Services*, Sustainable Development Goals Series,
https://doi.org/10.1007/978-981-19-3000-3_11

flora and fauna uniquely adapted to this surrounding. Yet, this globally valuable ecosystem is now vulnerable. This is due to the ever-increasing anthropogenic pressures (for example, over-harvesting, pollution, coastal development, destructive fishing and habitat degradation, climate change, intense and frequent natural disasters) buoyed by weak institutions and ineffective command-driven governance systems.

On a brighter note, this book demonstrates how customary sustainable practices and traditional knowledge of the Traditional Resource Users (TRUs) augment conservation through restoration and sustainable uses of biodiversity. However, this will only succeed if TRUs are given a chance and get supported by the government and non-government agencies. This argument is presented in this chapter and explained on the basis of a conceptual framework. A conceptual framework is an analytical tool with variations and contexts used to make conceptual distinctions and organise ideas. It can also be used in different categories of work where an overall picture is required. The discussion and results overall consist of (a) enhanced understanding of the value of mangrove for building resilience through bio-cultural diversity, (b) exploring area-based conservation and ecosystem-based solutions and (c) assisting the local coastal communities to become key contributors in making the ecosystem of the Sundarbans more resilient by ensuring that their rights are preserved over terrestrial, marine and coastal ecosystems and at the same time their services are better recognised and supported.

In addition to it, a multiple evidence-based approach for the assessment of the resilience capacity has been considered here. The findings of the assessment have been summarised by surveying the conceptual framework (developed through critical analysis of available secondary literature on natural resource management). This is done in conjunction with primary data gathered from the TRUs through several consultations and authors' appraisal of the situation. In this regard, a significant amount of primary data has been collected through participatory approaches such as Focus Group Discussions (FGDs), unstructured interview, Participatory Rural Appraisal (PRA) tools like social mapping, impact assessment done by the respondents particularly based on knowledge, perspectives and understandings of TRUs who are members of the three cooperatives under the Unnayan Onneshan.

Conceptual Framework: Informal Institutions, Sustainable Conservation and Resilience Capacity

A conceptual framework, as mentioned earlier, is an analytical instrument with variations and contexts that is used to draw conceptual distinctions and organise concepts. The framework, at its core, focuses on the means and measures employed for biodiversity resource management. The mainstream measures are derived mostly from market-centric theoretical underpinnings as part of the neoliberal intellectual agenda. According to this school of thought, biodiversity resource degradation is caused mostly by the absence of a market and negative externality (Perrings et al., 1992; Sandmo, 2015). It asserts that valuation techniques can provide useful insights to support policy initiatives by quantifying the economic value of resources and developing exchange rules in conjunction with biological resource protection (Barbier, 2007; Bräuer, 2003; Costanza et al., 1997; Hahn et al., 2015; Kumar, 2005; McAfee & Shapiro, 2010; Pearce, 2001). Institutional economists have complemented this understanding and they argue that a formal property rights regime can efficiently manage the natural resources whereas the absence of property rights results in resource degradation (Ituarte-Lima et al., 2014; Ostrom, 2000; Vatn, 2009, 2010).

A segment of the political economy analyses, on the other hand, perceives the presence of multiple property rights regimes as causing conflict in resource management and degradation. It exposes the hierarchical connection that exists in society and gives insight into the political features of the resource management system. It argues that institutional arrangements (property rights) are subject to certain political-economic factors resulting from the accumulation of wealth by different agents in presence of a non-cooperative solution. It also emphasises the responsibilities of formal political institutions and draws attention to narratives regarding ecosystem service changes (Robbins, 2012).

In Fig. 11.1, the political economy factors which induce biodiversity resource degradation is shown. The diagram reveals a series of events one after the other, which lead to the degradation of forest resources. It is evident that institutional fragility coupled with commercialisation leads to personalised gain accumulated by exerting political power that resulted in the degradation of the forest resources. The framework thus helps illustrate the bio-environmental interaction in the presence of a distribution of power to production activities and its connection to ecological analysis

(Greenberg & Park, 1994). It shows the importance of the claim that the degradation of natural resources is not only about the non-existence of the market but also about uneven power-sharing by the stakeholders over the management of resources. The prevalence of vertical ties in society as well as the upward enforcement of rules, make it easier for powerful groups to seize resources with relative ease (Adhikari & Goldey, 2010). The process puts the rule of individuals ahead of the rule of law, resulting in institutional fragility. Thus, powerful groups seize the rents and property rights.

Unfortunately, market-centric analysis fails to recognise the value of certain species thriving in a special kind of ecosystem and the fact that exploiting the species for satisfying the demand of the market may lead to its extinction. Hence, it falls short of providing a sustainable solution regarding the unique aspects of the interdependent linkages among humans, biodiversity resources and ecosystem services.

Economic valuation-based exchange has been proven to be flawed (Gómez-Baggethun & Ruiz-Pérez, 2011; Kosoy & Corbera, 2010; Muradian et al., 2013; Turnhout et al., 2013; Neuteleers & Engelen, 2015). It reduces biodiversity to a series of computable components, putting it to utilitarian use and reducing social-natural relationships to merely commercial transactions (Turnhout et al., 2013). Such measurements promote a narrow conceptualisation of ecosystem services and are likely to be severely detrimental to resource conservation. However, the political ecology tenet also does not propose any conservation measure, apart from providing a broad understanding of the contributing elements causing the degradation of resources.

Human beings are an integral part of the ecology; they are not exclusive from the system who are meant to extract resources only. The long-standing relation between human beings and ecology and the role they play remains unexplored. This is sometimes identified to be an external part of the system. But, humans have been maintaining an interwoven, intimate and reciprocal relationship with nature. This relation can be scrutinised from the standpoint of 'human sociality'. Human sociality refers to the human beings, as a collective organisation, and is part of the larger ecosystem, which possess distinct knowledge and practices that systematically and sustainably contributes to the conservation and regeneration of the resources along with maintaining provision of ecosystem services (Titumir et al., 2020). It emphasises the notion that societies that are in harmony with nature, contribute to biodiversity conservation via

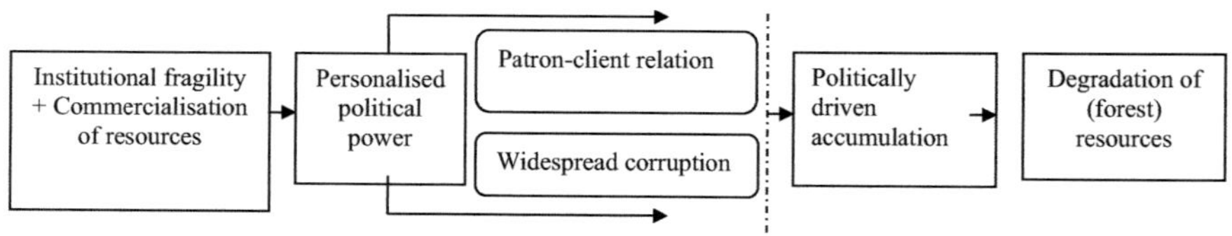

Fig. 11.1 Political economy factors inducing biodiversity resources degradation (*Source* Prepared by authors)

revitalisation where informal institution plays a crucial role. Informal institutions here include norms, values and traditional knowledge. It conserves and regenerates the resources for making a more resilient ecological system and society.

This chapter constructs a sustainable conservation framework that demonstrates that inter-institutional failure caused by the exclusion of informal institutions and community ownership promotes natural resource degradation. It also argues that the earlier practices of fencing off pieces of nature, as a means to 'mitigate' anthropogenic intervention, have been expensive, unsustainable and dubious in terms of socio-economic and conservation processes (Liu et al., 2012). Hence, the framework has used the political economics assumption to identify the causes of degradation, with an emphasis on the complementary relationship existing between humans and nature to ensure sustainable resource utilisation and distribution. It asserts that conservation necessitates the recognition of a range of values, knowledge and framing of socio-ecological production landscapes that establish cooperation and incentivise conservation for long-term sustainable use of those resources (Fig. 11.2).

In simpler words, Fig. 11.2 depicts the conceptual framework based upon both human sociality and patron-client relationship. It begins with portraying the patron-client relationships and the formal institutional arrangement based on it. This in turn branches out to depict the alienation of the TRUs and the developing institutional fragility. Both the outcomes depicted in the figure lead to unsustainable biodiversity of resource extraction, where the amount of resource harvested is more than what is being yielded. It branches out a third perspective which depicts the allocation of the TRUs. However, this point also lies under the conceptual framework based upon human sociality. In this part, the framework demonstrates an informal institution that branches out to the allocation of the TRUs as mentioned, along with their traditional knowledge and practices. These two outcomes lead towards sustainable utilisation and production of biodiversity resources, where the yield is greater than the amount harvested.

Alternatively, this framework contends that in the presence of neoliberal means and measures, the exchange process forms a patron-client relationship. Throughout this process, TRUs are relegated to becoming external agents in the ecological environment and it brings institutional fragility due to unequal power-sharing between political elites and TRUs.

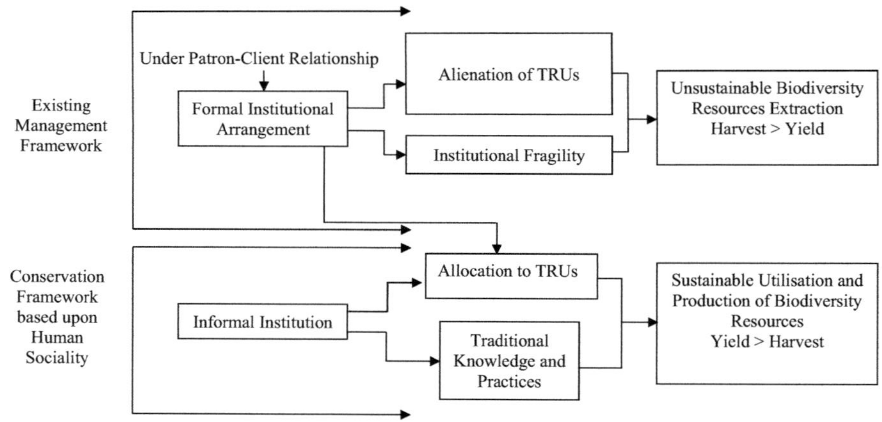

Fig. 11.2 Conceptual framework (*Source* Prepared by authors)

Such exchange arrangement leads to primitive accumulation and unsustainable resource exploitation (where the harvest is greater than the yield due to maximum realisation of the resources rent). On the contrary, the sustainable conservation framework based on human sociality implies that the allocation of resources regime to the TRUs is sustainable. TRUs with their traditional knowledge and practices form a socio-ecological production network. They contribute to maintaining this production network because of its symbiotic nature to the stock of resources. This incentivises TRUs to create knowledge to conserve the resources and to put the knowledge into practice for assuring a sustainable value chain. Thus, the TRUs and their traditional knowledge and practices make the biodiversity resources more resilient (where yield is greater than harvesting) and sustainable.

Assessment of the Resilience Capacity

This section seeks to assess the resilience capacity of the Sundarbans under two different scenarios. First, under the current conservation framework and second, under the alternative conservation framework which are developed in this chapter and the previous sub-section. The current management approach has already been discussed. Therefore, this section will provide a brief overview of the 'resilience indicators' as it has been chosen by the authors. Then, it will present the informal institutional arrangement that can be practised by the TRUs groups. These are also known as the key components of alternative conservation frameworks as argued in the previous section. Along with that, the resilience capacity will be determined based on those indicators.

Resilience Indicators for SEPLS

The resilience approach in the conceptual framework section proved to be useful while considering the potential to maintain, revitalise and rebuild a particular type of ecosystem. While discussing the profile and state of the forest, this book revealed that it can be classified as a perfect case of Socio-Ecological Production Landscapes and Seascapes (SEPLS) (please see chapter 2). Therefore, the study considers the 'resilience indicators of SEPLS'[1] to assess the case of the Sundarbans. Fundamental changes to SEPLS have the capability to unbalance customary sustainable use and processes, causing a decrease in resilience and increase in vulnerability. To avoid such a negative trend, it is crucial not only to obtain a better understanding of the 'components' of resilience but also to empower local communities and provide the necessary tools needed to understand this. Such indicators would create a strong foundation that may help to recognise negative trends and potential opportunities for further strengthening resilience. The following table exhibits the indicators and their components in detail (Table 11.1).

Table 11.1 shows the resilience indicators for SEPLS which consist of the resilience indicators along with some questions and the range of scores. The range of scores is technically the same for all the indicators mentioned. It ranges from a scale of 1 to 5 where 1 is considered to be 'very low'. This depicts that there are no opportunities for mobility. In contrast, a score of 5 is considered 'very high' and depicts that there are sufficient opportunities for mobility. In between these, a score of 2, 3 and 4 is considered to be 'low', 'medium' and 'high', respectively. The first few indicators consisted of landscape and seascape diversity and ecosystem protection.

The first point discusses landscape and seascape diversity; and questions whether or not the landscape and seascape are constituted of varied natural ecosystems (terrestrial and aquatic) and land uses. The second indicator talks about the ecosystem's protection and determines whether the regions within the landscape or seascape where ecosystems are protected are subject to formal or informal protection. Third, the indicator is concerned with the ecological interaction under consideration. This raises the question of whether ecological interactions between different components of the landscape or seascape were taken into account while managing natural resources. The fourth indicator is about recovery and regeneration, and it asks if the landscape or seascape can recover and regenerate after severe environmental shocks.

The next few indicators are concerned with biodiversity. Moving on, the fifth indicator is the diversity of the local food system. It questions whether or not the community consumes a range of locally produced

[1] The indicators are jointly developed by Biodiversity International and UNU-IAS.

Table 11.1 Resilience indicators for SEPLS

Resilience Indicators and Questions for Scoring	*Scores*
Landscape/seascape diversity and ecosystem protection	
1. Landscape and seascape diversity Is the landscape/seascape composed of diverse natural ecosystems (terrestrial and aquatic) and land uses?	(5) Very high (There is a large number of natural ecosystems and land uses) (4) High (3) Medium (2) Low (1) Very low (There is only one or a very small number of natural ecosystems(and land uses)
2. Ecosystem protection Are there areas in the landscape or seascape where ecosystems are protected under formal or informal forms of protection?	(5) Very high (Key resources are under some form of protection) (4) High (3) Medium (2) Low (1) Very low (There are no areas under protection)
3. Ecological interaction considered Are ecological interactions between different components of the landscape or seascape considered while managing natural resources?	(5) Very high (Ecological interactions are considered while managing natural resources) (4) High (3) Medium (2) Low (1) Very low (Ecological interactions are not considered while managing natural resources)
4. Recovery and regeneration Does the landscape or seascape have the ability to recover and regenerate after extreme environmental shocks?	(5) Very high (Very high ability to recover and regenerate) (4) High (3) Medium (2) Low (1) Very low (Very low ability to recover and regenerate)

Resilience Indicators and Questions for Scoring	*Scores*
Biodiversity	
5. Diversity of local food system Does the community consume a diversity of locally-produced food?	(5) Very high (Diversity of locally-sourced foods is very high and these foods are widely consumed) (4) High (3) Medium (2) Low (1) Very low (There are very few or no locally-sourced foods)
6. Maintenance and use of local crop varieties and animal breeds Are different local crops, varieties and animal breeds conserved and used in the community?	(5) Very high (Local crop varieties and animal breeds are widely conserved and used) (4) High (3) Medium (2) Low (1) Very low (There are few or no local crop varieties and animal breeds)
7. Sustainable management of biodiversity resources Are common resources managed sustainably?	(5) Very high (Common resources are managed sustainably) (4) High (3) Medium (2) Low (1) Very low (Common resources are overexploited or depleted)
Knowledge and innovation	
8. Innovation in agriculture and conservation practices Does the community develop, improve and adopt new agricultural, fisheries, forestry and conservation practices and/or revitalises traditional ones to adapt to changing conditions, including climate change?	(5) Very high (The community is receptive to change and adjusts its practices) (4) High (3) Medium (2) Low (1) Very low (The community is not receptive to change and makes few innovations)

(continued)

Table 11.1 (continued)

Resilience Indicators and Questions for Scoring	*Scores*
9. Traditional knowledge related to biodiversity Are local knowledge and cultural traditions related to biodiversity transmitted from elders and parents to young people in the community?	(5) Very high (Local knowledge and cultural traditions are transmitted to young people) (4) High (3) Medium (2) Low (1) Very low (Local knowledge and cultural traditions are lost)
10. Documentation of biodiversity-associated knowledge Is agricultural biodiversity and associated knowledge, documented and exchanged?	(5) Very high (Documentation is robust) (4) High (3) Medium (2) Low (1) Very low (There is little or no documentation in the community)
11. Women's knowledge Are women's knowledge, experiences and skills recognised and respected at household, community and landscape levels?	(5) Very high (Women's knowledge, experiences and skills recognized and respected at all levels) (4) High (3) Medium (2) Low (1) Very low (Women's knowledge, experiences and skills are not recognized and respected)
Governance and social equity	
12. Rights of the community in resource management Does the community have customary and/or formally recognised rights over land, (seasonal) pastures, water and natural resources?	(5) Very high (Rights are fully recognized and not disputed) (4) High (3) Medium (2) Low (1) Very low (Rights are not recognized and heavily disputed)

Resilience Indicators and Questions for Scoring	*Scores*
13. Community-based governance Is there a multi-stakeholder landscape/seascape platform or institution that is able to effectively plan and manage landscape resources?	(5) Very high (Platform or institution is capable of transparent, participatory and effective decision making) (4) High (3) Medium (2) Low (1) Very low (There is no multi-stakeholder platform or institution)
14.Social capital as cooperation and coordination in resource management Is there any connection, coordination and cooperation within and between communities for the management of natural resources?	(5) Very high (There is a very high level of cooperation and coordination in natural resource management) (4) High (3) Medium (2) Low (1) Very low (There is little or no cooperation and coordination in natural resource management)
15. Social equity Is access to opportunities and resources fair and equitable for all community members, including women, at household, community and landscape level?	(5) Very high (Access to resources and opportunities is fair and equitable at all levels) (4) High (3) Medium (2) Low (1) Very low (Access to resources and opportunities is not fair and equitable)
Livelihood and Well-being	
16. Socio-economic infrastructure Is the socio-economic infrastructure adequate for the needs of the community?	(5) Very high (Socio-economic infrastructure meets all community needs) (4) High (3) Medium (2) Low (1) Very low (Socio-economic infrastructure does not meet community needs)

(continued)

Table 11.1 (continued)

Resilience Indicators and Questions for Scoring	*Scores*
17. Human health and environmental conditions What is the general health situation of local people also considering the prevailing environmental conditions?	(5) Very high (Health situation and the environmental conditions are good) (4) High (3) Medium (2) Low (1) Very low (The health and the environmental conditions are bad)
18. Income diversity Are households in the community involved in a variety of sustainable, income-generating activities?	(5) Very high (Households are involved in a variety of sustainable, income generating activities) (4) High (3) Medium (2) Low (1) Very low (Households have no alternative economic activities)
19. Biodiversity-based livelihoods Does the community develop innovative use of the local biodiversity for its livelihoods?	(5) Very high (Livelihoods are being improved by innovative use of local biodiversity) (4) High (3) Medium (2) Low (1) Very low (Livelihood improvementsare not related to local biodiversity)
20. Socio-ecological mobility Are households and communities able to move around between different production activities and locations as necessary?	(5) Very high (There are sufficient opportunities for mobility) (4) High (3) Medium (2) Low (1) Very low (There are no opportunities for mobility)

Source UNU-IAS (2015)

food. The sixth item is about preserving and utilising local crop varieties and animal breeds. This point questions if different local crops, varieties and animal breeds are conserved and used in the community. The seventh indicator is about the sustainable management of biodiversity resources. This asks if common resources are managed sustainably. The next few indicators are concerned with knowledge and innovation. Moving on, the eighth indicator talks about the innovation in agriculture and conservation practices. It questions whether the community develops, improves and adopts new agricultural, fisheries, forestry and conservation practices and reinvents traditional ones to adapt to changing conditions which includes climate change. The ninth point talks about traditional knowledge related to biodiversity. It questions whether local knowledge and cultural traditions are related to biodiversity transmitted from elders and parents to young people in the community. The tenth point is about the documentation of biodiversity-associated knowledge. This point questions if agricultural biodiversity and associated knowledge are documented and exchanged. The eleventh point is concerned with women's knowledge and explores whether women's knowledge, experiences and skills are recognised and respected at the household, community and landscape levels.

The next following indicators are related to governance and social equity. The twelfth indicator talks about the rights of the community in resource management. It questions whether the community has customary and formally recognised rights over land, (seasonal) pastures, water and natural resources. The thirteenth indicator talks about community-based governance and explores whether there is a multi-stakeholder landscape and seascape platform or institution that is able to effectively plan and manage landscape resources. The fourteenth point is about social capital as cooperation and coordination in resource management. This point questions if there is any connection, coordination and cooperation between communities for the management of natural resources. The fifteenth point talks about social equity and questions whether access to opportunities and resources is fair and equitable for all community members, including women, at household, community and landscape level.

The last few indicators are related to livelihood and well-being. The sixteenth point talks about the socio-economic infrastructure and questions whether the socio-economic infrastructure is adequate for serving the needs of the community. The seventeenth point is about human health and environmental conditions. This questions what is the general health situation of the local people alongside the prevailing environmental conditions. The eighteenth point is about income diversity and it asks whether households within the community are involved in a variety of sustainable income-generating activities. The nineteenth indicator is about biodiversity-based livelihoods. This questions whether the community develops innovative use of the local biodiversity for its livelihoods. Finally, the twentieth indicator is socio-ecological mobility. This indicator questions whether households and communities are able to move around between different production activities and locations as necessary.

Enhancement of Resilience Under Informal Institutions-Based Conservation Framework

The study incorporates twelve indicators out of twenty in this case. A comparison of two conservation approaches signifies that informal institutions-based alternative framework contributes significantly to the conservation of the biodiversity by creating a more resilient ecology and society (Table 11.2). This conservation approach directly puts an impact on 12 resilient indicators by indicating a positive relationship. It denotes that this framework is more ecologically responsive to the context of SEPLS. For instance, under the present management approach, the ecosystem is minimally protected and the regeneration capacity is hindered. This happens due to the inability to check anthropogenic pressures. On the contrary, the alternative framework attempts to ensure the protection of the ecosystem at a greater level and stimulates the regeneration capacity at the fullest (indicators 1, 2 and 3). This is conceivable because the alternative regime puts a greater emphasis on the importance of the traditional knowledge system in SEPLS management, whereas the existing regime does not fully recognise traditional knowledge (indicators 5 and 6). In actuality, community-based governance is only conceptualised in the policy document in terms of the governance and equity indicators. Such governance systems are undermined by agencies of the government themselves. Hence, the alternative suggestions include the participation of the community build a social capital that contributes to cooperation, social equity and efficient governance (indicators 7, 8, 9 and 10). In fact, such a framework seeks to uphold the biodiversity based on livelihood patterns (therefore, there will be no change in this indicator) that is present in that mangrove ecosystem. However, it argues that this

Table 11.2 Comparative analysis of resilience capacity of the Sundarbans under two different scenarios

Resilience Indicators	*Scenario under Current Practice of Management*					*Scenario under the Alternative Conservation Framework*				
	Very High	*High*	*Medium*	*Low*	*Very Low*	*Very High*	*High*	*Medium*	*Low*	*Very Low*
Landscape and seascape diversity and ecosystem protection										
Ecosystem protection				√			√			
Ecological interaction considered				√		√				
Recovery and regeneration					√	√				
Biodiversity										
Sustainable management of biodiversity resources					√	√				
Knowledge and innovation										
Traditional knowledge related to biodiversity			√			√				
Documentation of biodiversity-associated knowledge					√	√				
Governance and social equity										
Rights of the community in resource management				√		√				
Community-based governance				√		√				

(continued)

Table 11.2 (continued)

Resilience Indicators	*Scenario under Current Practice of Management*					*Scenario under the Alternative Conservation Framework*				
	Very High	*High*	*Medium*	*Low*	*Very Low*	*Very High*	*High*	*Medium*	*Low*	*Very Low*
Social capital as cooperation and coordination in resource management					√	√				
Social equity					√		√			
Livelihood and well-being										
Income diversity				√			√			
Biodiversity-based livelihoods	√					√				

Source Prepared by the authors

should be acted upon in a sustainable manner, so that it conserves the biodiversity resources (indicator 4) and provides alternative livelihoods to the humans, under the changed circumstances by diversifying their income sources (indicator 11).

In other words, Table 11.2 shows the comparative analysis of resilience capacity under two different scenarios, that is under current management and under alternative conservation framework. The indicators are scored as per their ratings under both these different scenarios in the table. Ecosystem protection shows a low score under current management practice and a high score under the alternative conservation framework. Ecological interaction shows a low score under current management practice and a very high score under the alternative conservation framework. Recovery and regeneration show a very low score under current management practice but a very high score under the alternative conservation framework. Moving on, sustainable management of biodiversity resources show a very high score under the current management practice but a very low score under alternative conservation framework. Traditional knowledge related to biodiversity shows a medium score under current management practice but a very high score under alternative conservation framework. Documentation of biodiversity-associated knowledge shows a very low score under current management practice but a very high score under alternative conservation framework. Rights of the community in resource management show a low score under current management practice but a very high score under alternative conservation framework. Community-based governance shows a low score under current management practice but a very high score under alternative conservation framework. Social capital as cooperation and coordination in resource management shows a very low score under current management practice but a very high score under alternative conservation framework. Social equity shows a very low score under current management process but a high score under alternative conservation framework. Income diversity shows a low score under current management process and a high score under alternative conservation framework. Lastly, biodiversity-based livelihoods show a very high score under both current management practise as well as alternative conservation frameworks.

Aichi Biodiversity Targets and SDGs[2] Under Alternative Conservation Framework

The alternative conservation framework also helps in obtaining some of the essential targets outlined in the 'Aichi Biodiversity Targets'[3] as illustrated in Table 11.3. The table depicts three distinct targets as per the indicators, along with their relevant issues and how those factors contribute to this case study. First, Target no. 10 'Pressures on vulnerable ecosystems reduced' identifies pertinent concerns such as changes in the size of vulnerable ecosystems (here mangrove), anthropogenic pressures and climate change. The contribution this factor made was it had multiple anthropogenic pressures identified on a mangrove ecosystem and it presented and promoted the TK-based climate adaptation methods and sustainable agricultural methods. Second, Target no. 15 'Ecosystem restored and resilience enhanced' had its relevant issues as ecosystem resilience and restoration. The contribution this point had was, it made traditional rules and methods followed by TRUs enhancing the restoration process and improving resilience capacity. Also, it had climate change adaptation approaches such as CMAAS innovated by the TRUs enhancing resilience capacity. Third, Target no. 18 'Traditional knowledge respected' had its relevant issues such as traditional knowledge, innovations and practices, along with customary use of biological resources. The contributions these had were, promoted TK knowledge system practised by the TRUs. Then, it urged to recognise the traditional practices in the resource management framework. Last, it articulates the involvement of TRUs in the resource management.

In short, it helped to contribute to Target no. 10 by lowering the pressure on vulnerable (here, mangrove) ecosystems. Then, it promoted revival and improved the resilience of that ecosystem and thus helped to achieve Target no. 15. Most importantly, it devoted to achieve Target no. 18 by respecting the TK system which was practised by the local and indigenous communities (Table 11.3).

[2] The United Nations' Sustainable Development Goals (SDGs) are a plan for a more sustainable future for everyone. There are 17 goals and 169 targets under the umbrella of SDG.

[3] Aichi Biodiversity Targets- In decision X/2, the 10th meeting (2010) of the Conference of the Parties for the Convention on Biological Diversity (CBD), adopted a revised

Table 11.3 Achievement of Aichi Biodiversity Targets under the alternative conservation framework

Targets	*Relevant Indicators/Issues*	*Contribution of this Study*
Target 10: *Pressures on vulnerable ecosystems reduced*	• Trends in the extent of vulnerable ecosystems (here mangrove) • Anthropogenic pressures • Climate change	• Multiple anthropogenic pressures identified on a mangrove ecosystem • Presenting and promoting the TK based climate adaptation methods and sustainable agricultural methods
Target 15: *Ecosystem restored and resilience enhanced*	• Ecosystem resilience • Restoration	• Traditional rules and methods followed by TRUs enhances the restoration process and improves resilience capacity • Climate change adaptation approaches like CMAAS innovated by the TRUs enhances resilience capacity
Target 18: *Traditional knowledge respected*	• Traditional knowledge, innovations and practices • Customary use of biological resources	• Promotes TK knowledge system practised by the TRUs • Urges to recognise the traditional practices in the resource management framework • Articulates the involvement of TRUs in the resource management

Finally, conservation framework incorporating the TRUs predominantly can help achieve SDG 14 and 15. Furthermore, some of the targets under SDG 6 (Target 6.6), SDG 11 (Target 11.4), SDG 12 (Target 12.2) and SDG 13 (Target 13.3) can also be attained indirectly (Table 11.4).

and updated Strategic Plan for Biodiversity, including Aichi Biodiversity Targets for the 2011–2020 period. It has five strategic goals.

Table 11.4 SDGs under the alternative conservation framework

SDGs	*Target(s)*
Major implications	
SDG 14: Life below water	Target 14.2: Protect and restore ecosystems Target 14.4: Sustainable fishing Target 14.5: Conserve coastal and marine areas
SDG 15: Life on land	Target 15.1: Conserve and restore terrestrial and freshwater ecosystems Target 15.2: End deforestation and restore degraded forests Target 15.5: Protect biodiversity and natural habitats
Indirect implications	
SDG 6: Clean water and sanitation	Target 6.6: Protect and restore water-related ecosystems
SDG 11: Sustainable cities and communities	Target 11.4: Protect the world's cultural and natural heritage
SDG 12: Responsible consumption and production	Target 12.2: Sustainable management and use of natural resources
SDG 13: Climate action	Target 13.3: Build knowledge and capacity to meet climate change

Source Prepared by authors

Conclusions

This chapter helps to bring out the resilience factors and contributions of the Traditional Resource Users (TRUs) towards the SDGs and Aichi Biodiversity Targets. It also helped to reveal that, if the TRUs get the support of the government and non-government agencies, they will be able to contribute a lot towards the forest and assist in maintaining its valuable resources and biodiversity. The conceptual framework in this chapter incorporates informal institutions, sustainable conservation and resilience capacity in SEPLS management. The political economy factors inducing biodiversity resource degradation was also an outcome of institutional fragility coupled with commercialisation. It helped to expose that the degradation of natural resources was not only due to the non-existence of the market but also due to the distorted power-sharing between the stakeholders over the management of resources. Furthermore, it was seen how the conceptual framework that took into account patron-client relation and conservation framework based upon human sociality played its roles. It revealed that the conservation framework based on human sociality led to a better outcome. This is because it led to the sustainable utilisation and production of biodiversity resources, where the amount that was yielded was greater than the amount being harvested. This was in contrast to the outcomes under the patron–client relationship which resulted in an unsustainable biodiversity resource extraction that led to a greater amount of harvest compared to a relatively lower amount of yield. To sum it up, this acted as an incentive to conserve the knowledge and ensure a sustainable value chain which will boost a good practise and help maintain resources.

In addition, this chapter also helped to reveal the assessment of the resilience capacity under two different types of the framework that is current conservation framework and the alternative conservation framework. The resilience indicators were listed down along with questions for scoring. The assessment of the forest based on the resilience indicators of SEPLS also depicted that the current resilience capacity can be developed by generalising the traditional knowledge base and involving the indigenous people into the resource management framework. It is evident that there ought to be clear proximity between traditional knowledge and the formal conservation stream. Thus, the SEPLS indicators need to be analysed. Local communities and formal authorities should make a collaborative effort to serve the purpose.

The results from the analysis were such that nine out of the twelve indicators that were listed showed very high scores and the rest three indicators showed high scores for the alternative conservation framework. This was in comparison to the current management practice where most of the indicators had low or very low scores. Therefore, it can be said from this analysis that the alternative conservation framework allows sufficient opportunity for mobility, as per the high and very high scores; whereas the current management process depicts mostly no opportunities for mobility, as per the low and very low scores. The chapter also revealed the achievements of the Aichi Biodiversity Targets which also comes under the alternative conservation framework. This had targeted three points from the indicators and demonstrated their relevant issues in addition to the contributions in this case, through various methods. Some of these were the promotions of traditional knowledge-based climate adaptation methods, traditional rules and methods followed by TRUs, etc. The chapter ends by mentioning the specific SDGs and associated targets that can also be attained under the alternative conservation strategy.

Acknowledgements The chapter draws on the authors' previous publication of a book chapter in 2020 entitled—'Traditional knowledge, institutions and human sociality in sustainable use and conservation of biodiversity of the Sundarbans of Bangladesh', in Managing Socio-ecological Production Landscapes and Seascapes for Sustainable Communities in Asia: Mapping and Navigating Stakholders, Policy and Action, eds. O. Saito, S.M. Subramanian, H. Hashimoto & K. Takeuchi, Science for Sustainable Societies, pp. 67–92.

References

Adhikari, K. P., & Goldey, P. (2010). Social capital and its "downside": The impact on sustainability of induced community-based organizations in Nepal. *World Development, 38*(2), 184–194.

Barbier, E. B. (2007). Valuing ecosystem services as productive inputs. *Economic Policy, 22*, 177–229.

Bräuer, I. (2003). Money as an indicator: To make use of economic evaluation for biodiversity conservation. *Agriculture, Ecosystems and Environment, 98*, 483–491.

Costanza, R., D'Arge, R., de Groot, R., Farberk, S., Grasso, M., Hannon, B., Limburg, K., Naeem, S., O'Neill, R. V., Paruelo, J., Raskin, R. G., Sutton, P., & van den Belt, M. (1997). The value of the world's ecosystem services and natural capital. *Nature, 387*, 253–260.

Gómez-Baggethun, E., & Ruiz-Pérez, M. (2011). Economic valuation and the commodification of ecosystem services. *Progress in Physical Geography, 35*, 613–628. https://doi.org/10.1177/0309133311421708

Greenberg, J. B., & Park, T. K. (1994). Political ecology. *Journal of Political Ecology, 1*, 1–12.

Hahn, T., McDermott, C., Ituarte-Lima, C., Schultz, M., Green, T., & Tuvendal, M. (2015). Purposes and degrees of commodification: Economic instruments for biodiversity and ecosystem services need not rely on markets or monetary valuation. *Ecosystem Services, 16*, 74–82.

Ituarte-Lima, C., Schultz, M., Hahn, M.C., & Cornell, S. (2014). *Biodiversity financing and safeguards: Lessons learned and proposed guidelines*, Stockholm: SwedBio/Stockholm Resilience Centre at Stockholm University, Information Document UNEP/CBD/COP/12/INF/27 for the 12th Conference of the Parties of the Convention on Biological Diversity in Pyeongchang Korea.

Kosoy, N., & Corbera, E. (2010). Payments for ecosystem services as commodity fetishism. *Ecological Economics, 69*, 1228–1236.

Kumar, P. (2005). *Market for ecosystem services*. International Institute for Sustainable Development (IISD).

Liu, J. J., Harris, L., Zhao, H. J., & Qian, F. (2012). Integrating community development with the management of grasslands at Ke'erqin Nature Reserve, Inner Mogolia, China. *IPSI Secretariat, IPSI case study booklet*. United Nations University Institute for the Advanced Study of Sustainability.

McAfee, K., & Shapiro, E. N. (2010). Payments for ecosystem services in Mexico: Nature, neoliberalism, social movements, and the state. *Annals of the Association of American Geographers, 100*, 579–599.

Muradian, R., Arsel, M., Pellegrini, L., Adaman, F., Aguilar, B., Agarwal, B., Corbera, E., Ezzine de Blas, D., Farley, J., & Froger, G. (2013). Payments for ecosystem services and the fatal attraction of win-win solutions. *Conservation Letters, 6*, 274–279.

Neuteleers, S., & Engelen, B. (2015). Talking money: How market based valuation can undermine environmental protection. *Ecological Economics, 117*, 253–260.

Ostrom, E. (2000). *Private and common property rights: Population and environmental change*. Indiana University.

Pearce, D. (2001). Valuing biological diversity: Issues and overview. In *Valuation of biodiversity benefits: Selected studies*. OECD.

Perrings, C., Folke, C., & Mäler, K. (1992). The ecology and economics of biodiversity loss: The research agenda. *Ambio, 21*(3), 201–211.

Robbins, P. (2012). *Political ecology: A critical introduction* (2nd ed.). Blackwell.

Sandmo, A. (2015). The early history of environmental economics. *Review of Environmental Economics and Policy, 9*(1), 43–63. https://doi.org/10.1093/reep/reu018

Titumir, R. A. M., Afrin, T., & Islam, M. S. (2020). Traditional knowledge, institutions, and human sociality in sustainable use and conservation of biodiversity of the Sundarbans of Bangladesh. In O. Saito, S. M. Subramanian, S. Hashimoto, & K. Takeuchi (Eds.), *Mapping socio-ecological production landscapes, stakeholders, policy and action*. Science for Sustainable Societies, Integrated Research System for Sustainability Science (IR3S), University of Tokyo. https://doi.org/10.1007/978-981-15-1133-2_5

Turnhout, E., Waterton, C., Neves, K., & Buizer, M. (2013). Rethinking biodiversity: From goods and services to "living with." *Conservation Letters, 6*, 154–161.

UNU-IAS. (2015). *Indicators of resilience in socio-ecological production landscapes and seascapes* (2nd ed.). Tokyo, Japan.

Vatn, A. (2009). An institutional analysis of methods for environmental appraisal. *Ecological Economics, 68*, 2207–2215.

Vatn, A. (2010). An institutional analysis of payments for environmental services. *Ecological Economics, 69*, 1245–1252.

CHAPTER 12

Human-Nature Cooperation for Well-Being: Community Understanding on 'One Health Approach' in COVID-19 Era

Rashed Al Mahmud Titumir and *Md. Shah Paran*

Introduction

This chapter explores the interrelationship and cooperation between human beings and nature that directly and indirectly affect the quality of life for both and conjoins participatory research approaches to ascertain the community understanding of 'one health' approach that includes human, animal and ecosystem health during COVID-19 outbreak era in the Sundarbans. It examines the state of TRUs' health and well-being in terms of universal access to health, education and social security programmes including food security, job security, social security as

R. A. M. Titumir (✉)
Department of Development Studies, University of Dhaka, Dhaka, Bangladesh
e-mail: rt@du.ac.bd; rtitumir@unnayan.org

R. A. M. Titumir · Md. S. Paran
Unnayan Onneshan, Dhaka, Bangladesh

R. A. M. Titumir (ed.), *Sundarbans and its Ecosystem Services*, Sustainable Development Goals Series,
https://doi.org/10.1007/978-981-19-3000-3_12

well as the state of ecosystem' health in terms of biodiversity, anthropogenic pressures, resource management, sustainable production and consumption in the face of COVID-19. Accordingly, the study categorises human's contribution to nature (biodiverse adaptation to climate change, nature-based community solution for livelihood diversification, customary sustainable practices based on traditional knowledge) and nature's contribution to human beings (multiple ecosystem services) in promoting 'one health' in the Socio-Ecological Production Landscapes and Seascapes (SEPLS) in the Sundarbans.

The Sundarbans of Bangladesh, known as the lungs of the country, is a vital case of an ecologically vulnerable ecosystem with respect to biodiversity degradation (Titumir et al., 2020). Over the last few decades, it has been experiencing major ecological and physiographical changes and losing its resources due to both human interventions and climatic changes. The area of Sundarbans in Bangladesh part was 17,000 km^2 in 1776, which has reduced to almost half of what it was previously (Islam & Gnauck, 2009). Moreover, the outbreak of the COVID-19 pandemic has posed unprecedented challenges to people's life and livelihood around the world. The forest people in the Sundarbans are also facing the brunt of the pandemic. Around 95% TRUs in the Sundarbans have incurred huge losses in income during nationwide lockdown (Unnayan Onneshan, 2020c). The socio-economic distress of the forest people is negatively affecting the forest by creating an imbalance in the ecosystem. To top this, the coastal part of Bangladesh was ravaged by super cyclone *Amphan* on 16 May 2020 which resulted in damaged crops, infrastructures and coastal protection embankment in 26 coastal districts, accentuating the livelihood crisis of the Indigenous People and Local Communities (IPLCs) in the Sundarbans.

Meanwhile, due to the absence of universal social security programmes in the country, the economic fallout for the majority of the people has aggravated (Unnayan Onneshan, 2020b). The ongoing health crisis and livelihood insecurity resulting from COVID-19 has further marginalised the forest people and is also creating a metabolic rift which is a disruption induced in the ecosystem balance through breaking the producer, consumer and decomposer cycle. This imbalance in the ecosystem cycle has led to the deplorable condition of the forest and its people.

The ecosystem provides food and health security, clean air and water and livelihoods in form of multiple goods and services which contribute to good quality of life such as the physical and mental well-being of human beings (IPBES, 2019). At the same time, human beings as a part of the ecosystem can significantly contribute towards maintaining the ecosystem cycle. However, various anthropogenic pressures such as deforestation, agricultural intensification, wildlife exploitation, mining, infrastructure development, etc. are disrupting the ecosystem balance which results in the outbreak of zoonotic diseases like COVID-19 (IPBES, 2019). Maintaining ecosystem balance, therefore, requires a holistic approach that can ensure healthy, secure and good quality of life for human beings, animals and the ecosystem simultaneously. This study is an attempt to understand the 'one health approach' from a community perspective in the COVID-19 era as well as to inquire about the human–nature cooperation in the Sundarbans.

Forest and Impact of COVID-19 on TRUs

This globally acclaimed heritage site, including its sanctuaries and ecologically critical areas, which also acts as a natural wall to climatic exigencies, for example, cyclones (see Chapter 8), is vulnerable due to anthropogenic pressures and effective governance system. The ill-being of the forest also negatively affects the life and livelihood of its people who are further distressed due to the impact of COVID-19.

On a positive note, the customary sustainable practices, indigenous and local knowledge relating to conservation, restoration, sustainable uses and access and benefit-sharing may lead to transformative pathways, if such pathways are recognised by agencies concerned (see Chapter 4 & 5). The decadal changes of forest coverage in a part of the forest falling under Khulna administrative range, drawn by PPGIS, indicates that the number of trees in the forest is declining drastically. As a result, the amount of fallow land is increasing. In two decades, the total area of dense forests in the area of this study has halved (see Chapters 2 and 3).

Impact of COVID-19 on TRUs

The TRUs, the indigenous and local communities living in and depending on the forest for their livelihoods, historically face multiple pressures due to clientelist systems, functioning in the forest who use the management for marginalising, discriminating and disadvantaging them as customary owners and users. Increasing climatic impacts such as floods, cyclones and salinity intrusion are also hitting these communities hard. The frequency of cyclones on the coast of Bangladesh is increasing by the day damaging mangroves and the communities are vulnerable to the impacts. Houses, equipment and livelihood investments are frequently being destroyed (see Chapter 8).

In this context of existing hardships, the COVID-19 pandemic, nationwide lockdown, and the super cyclone *Amphan* have posed catastrophic impacts on the lives of millions of forest people. Approximately 95 per cent of the TRUs in the Sundarbans have incurred huge losses in terms of income and 51% of them have no jobs. The ongoing crisis will increase the poverty rate among the forest people to 40%, and 43.5% of the households' income is below the international poverty lines (Unnayan Onneshan, 2020c).

Khalil Dhali, a 53-year-old TRU and the president of Koyra Forest People Cooperative in the Sundarbans, has been collecting resources from the forest for years. The super cyclone Amphan devastated the livelihoods of millions like Dhali, leaving wreckages in the forest. He detailed the aftermaths of Amphan saying,

> Amphan destroyed the resources of the forest. It has swept away our home, damaged the embankments, inundated farmlands with saline water that become unfit for further cultivation. The water of ponds and tube wells has become saline too. We have nothing left. We cannot even collect resources as before. Many of us are starving. There is no option left for switching the occupation currently due to coronavirus.

The nationwide lockdown has caused a severe fall in demand for resources collected by the TRUs in the Sundarbans. They are getting a lower price for their resources in the market. At times they even fail to go to the markets for several months due to lockdown and social distancing measures in place which has left most of them with no income at all. Amori Begum, a female TRU, who usually goes to the forest, collects resources and contributes to her family like her male counterpart, said:

> We had to sit idle during the lockdown. We had not been allowed to go to the forest and to the markets. We had no income for several months. Amphan caused another catastrophic blow on our livelihoods at a time. Even after lockdown has been relaxed, we are not getting proper prices of the resources in the market. People are less willing to buy than before.

The forest department has imposed a ban on fishing from all rivers and canals in the Sundarbans for two months from 1 July to 30 August as part of the regular recommendations of the Integrated Resources Management Plans (IRMP) made as per an agreement in 2019 to protect the fishery resources. This ban has cropped up another adverse impact on the livelihood of TRUs who have already in peril due to COVID-19 and cyclone *Amphan*. Khalil Dhali adds,

> We have been sitting jobless for another two months. But we have to pay interest to money lenders. How can we pay interest of borrowed loans without an income? A similar ban was imposed last year; however, the situation is different this year due to coronavirus and Amphan. We cannot engage in other occupations due to coronavirus and we have not even received any compensation or incentives for the ban amid the ongoing crisis.

Bisaka Munda, an indigenous female TRU in the Sundarbans, has been collecting *golpata* (*Nypa fruticans*), catching fish and crabs from the forest and selling them to the local market. With her income, she has been significantly contributing to her family. When the nationwide lockdown was imposed in response to the COVID-19 pandemic, she had to sit idle and was not allowed to go to the forest or the markets. As a result, Bisaka, like many other TRUs, incurred huge losses. To add to this ongoing crisis, the coastal region of Bangladesh had to face the brunt of, *Amphan*, a super cyclone, which damaged her house and crop, further aggravating her crisis along with many other indigenous and local people. Her husband, also a TRU dependent on the forest, lost his income and was unable to find another job due to a lack of alternative options. The economic hardship resulting from COVID-19 and frequent natural disasters often increase domestic violence in the families like that of Bisaka's.

Disruption in the Ecosystem

According to the TRUs, the expropriation of resources, through over-harvesting and numerous anthropogenic pressures, are disrupting the ecosystem balance triggering massive biodiversity degradation in the forest (Fig. 12.1). As a consequence, the amount and quality of ecosystem services has deteriorated which is leading to livelihood distress for the

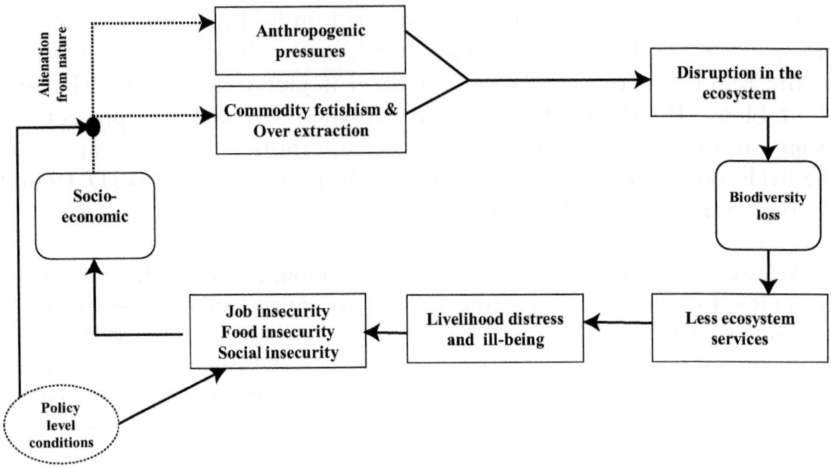

Fig. 12.1 State of the forest and its people: Biodiversity loss and livelihood distress (*Source* Adapted from Titumir & Paran, 2022a)

TRUs in the forest. Furthermore, according to TRUs, the policy regime is failing to secure job, food and social security for the people adequately. Hence, the existing socio-economic distress leads to dire suffering for them which negatively affects the ecosystem posing further pressures (Fig. 12.1).

The expropriation of the forest continues, according to them, it is when people extract more resources unsustainably for getting more money in the market. It can be said that the commodification of resources is derived from the alienation of human being from nature that heightens when people do not consider the true intrinsic value of nature (Titumir et al., 2019).

Policy Regime and Response to COVID Impact

The current policy response in Bangladesh has been found inadequate to the needs of the majority of the people—the informal sector workers, poor, vulnerable, lower-middle class, middle class and other disadvantaged portion of the people—in face of the COVID-19. The healthcare sector of the country is caught in a quagmire crumbling under disease burden for several months. Doctors, nurses and technicians are busy dealing with the newly emerged pandemic whereas, the non-communicable diseases are not being handled with optimum attention. To date, the healthcare sector comprises only 0.9% of GDP despite constant suggestions provided by organisations such as the WHO (Unnayan Onneshan, 2020b). The WHO recommends that the health budget should occupy around 4% of the GDP which is not followed in reality (Unnayan Onneshan, 2020b).

Budget allocation for education and technology is not up to the mark yet. This sector is going to get BDT 85,760 crores in the FY2020–21 budget which is 15.1% of the total budget (Unnayan Onneshan, 2020b). The education and technology sector takes 2.7% of the GDP which is not sufficient at all (Unnayan Onneshan, 2020b). The ongoing education system has been failing to generate skill development, innovation and building citizenship capacities in the country (Unnayan Onneshan, 2020b). On the other hand, there is no provision of universal social security programmes in the country. The ongoing targeted approach—social safety net programmes has not been able to curb the fallout from shocks—particularly COVID-19. Existing social protection programmes are inadequate and fragmented. The selection of beneficiaries is also mired by exclusion and inclusion errors (Unnayan Onneshan, 2020b). The stimulus package announced as an immediate response to the COVID-19 has also been found to be ineffective for majority of the people. The powerful and clientelist syndicate are grabbing the opportunities while the poor, vulnerable and disadvantaged people are being marginalised. The lack of adequate and effective response from the government has heightened the sufferings of the IPLCs in the Sundarbans as well. The livelihood insecurity has plunged the forest people into unprecedented precarity.

Community Conceptualisation of 'One Health Approach' in the Sundarbans

The TRUs consider the Sundarbans to be their mind which means they equate their life and well-being with the life and well-being of the forest as well as their thinking process and way of life around the life and spirit of the forest (see Titumir et al., 2019; Titumir & Paran, 2022a, b, c). The forest looks into the people's existence, livelihoods, social harmony, the breeding of their offspring, natural safety and security, and well-being. The IPLCs in the Sundarbans count on it. Their lives are influenced by the plethora of amenities offered by this forest which combines numerous values and contributes to well-being. The well-being, as they understand, is the healthy and secure life for both the forest and themselves attained by maintaining ecosystem balance. Therefore, the TRUs signify both the biotic and abiotic features of the ecosystem to realise a good quality of life. According to them, if the forest and its biodiversity remain safe and sound, they can also have a healthy and good quality of life. Securing healthy life for humans, animals and the ecosystem simultaneously, as they signify, is the mainstay to realise well-being in the ecosystem. This understanding leads them to contribute towards conservation, preservation and restoration of nature, which leads to the well-being of the ecosystem (Fig. 12.2).

Therefore, the interdependent relationship between IPLCs and the Sundarbans ascertains the term living in harmony with nature (Fig. 12.2). In contrast, according to them, if the alienation of human being from nature prevails, it leads to commodification and thus massive extraction of resources which result in disruption of the ecosystem and biodiversity loss (Fig. 12.3).

The TRUs say, human beings often consider themselves 'independent' and 'master' of nature, in spite of being a part of the ecosystem. They

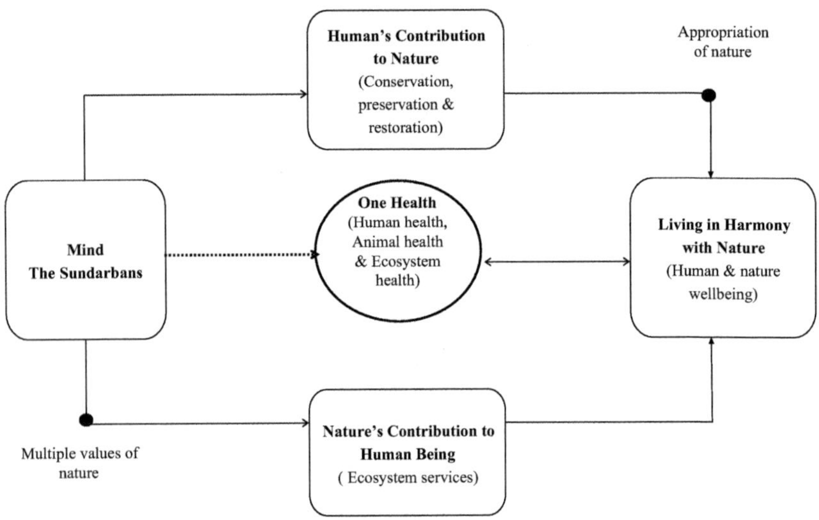

Fig. 12.2 Community conceptualisation of one health approach (*Source* Adapted from Titumir & Paran, 2022a)

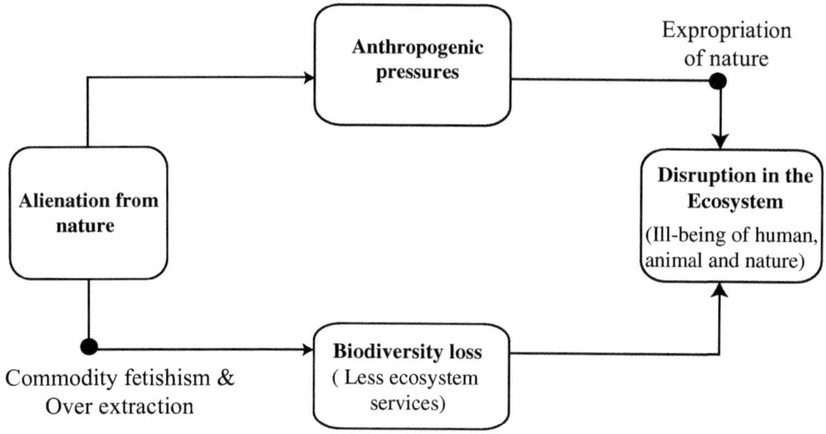

Fig. 12.3 Alienation from nature and biodiversity loss: ill-being (*Source* Adapted from Titumir & Paran, 2022a)

argue, the alienation from nature, treating nature as 'mere matter' or 'asset class', leads to over-extraction and destruction of nature in various ways. Therefore, human beings become separated from the nature when they fail to understand the true value of it and are unable to find themselves as part of the ecosystem and recognise the sociality between human and nature (Titumir et al., 2019). This alienation, as they suggest, results in the commodification of the natural resources through market pricing. In other words, valuation of nature becomes equal to the market price which causes over-extraction, thus, destruction of the natural resources (Fig. 12.3).

The alienation of human being from nature derived from the materialistic provision of commodities push the capitalist and powerful groups to get access to the resources in an unsustainable manner (Titumir et al., 2019). According to the TRUs, powerful groups are confiscating the resources through illegal means that aggravate due to extractive nature of institutions. As a result, the expropriation of nature continues with greater magnitude and as a consequence, TRUs and the primary producers become deprived of equitable benefit-sharing of natural resources.

Nature's Contribution to Human Beings

The TRUs identify multiple values that they get from the Sundarbans in the forms of numerous ecosystem services (ES). They get food, fuel and water as direct use-value. The Sundarbans also provide climate regulation, cyclone protection and erosion control as indirect use-value. The option value derives from the purview of the forest in research, education and aesthetics. Besides, the forest act as a cultural heritage, historical legacy with biodiversity, habitat, etc. which constitutes option value and existence value of the forest.

The forest provides livelihoods, enjoyment and recreation and serves as the protecting belt of the coastal region. The TRUs associate this forest as their lives on which their existence depends. They identify multiple ES that contribute to their well-being. These services include provisioning (foods, woods, fuel and water), supporting (nutrient cycle, soil formation and primary production), regulating (climate and flood regulation, disease prevention and water purification) and cultural (aesthetic, spiritual, educational and recreational) (see Chapter 6 & 7).

There are, however, many factors that affect the quality and amount of ecosystem services. When these factors are altered, the provision of well-being also become affected. These factors can be categorised in three different conditions namely the policy level or institutional conditions, the production level conditions and the ecological conditions. The policy level conditions are institutional and governance system, power and class, property rights and legal arrangements. If these factors function inclusively, they positively affect the ecosystem's health. In contrast, if policy level conditions take extractive form, it causes disruption in the ecosystem and thus harm the human, animal and ecosystem. The production level conditions include the level of carbon emission and pollution, the rate of extraction, degradation and exclusion, the nature of harvesting and fishing, the rate of reforestation and innovative use, etc. If production and consumption move forward in a sustainable and green manner, the quality of the ecosystem remains healthy. Otherwise, anthropogenic pressures will lead to massive depletion of nature and thus harm all biotic and abiotic features of the ecosystem (Titumir et al., 2022). The ecological conditions are climate change, weather patterns, natural disasters and hazards which also determine the provision of ES and the level of well-being of the ecosystem (see Chapter 6 for more).

Human's Contribution to Nature

In response to the continuous biodiversity loss negatively affecting the well-being of both the forest and its people, the community members have adopted several innovative practices for biodiversity restoration and conservation. They have started plantation programmes, nature-based production of resources such as crabs, fish, *golpata* (*Nypa fruticans*), etc. and also have been working as a watchdog to monitor the biodiversity status in the forest. Their activities for the appropriation of nature contribute to the well-being of nature (Fig. 12.4).

Promotion of Customary Sustainable Practices and Traditional Knowledge

The members of the cooperatives have developed specific sustainable practices following traditional and customary knowledge which are resilient and adaptive to climate change and can be promoted as innovative models

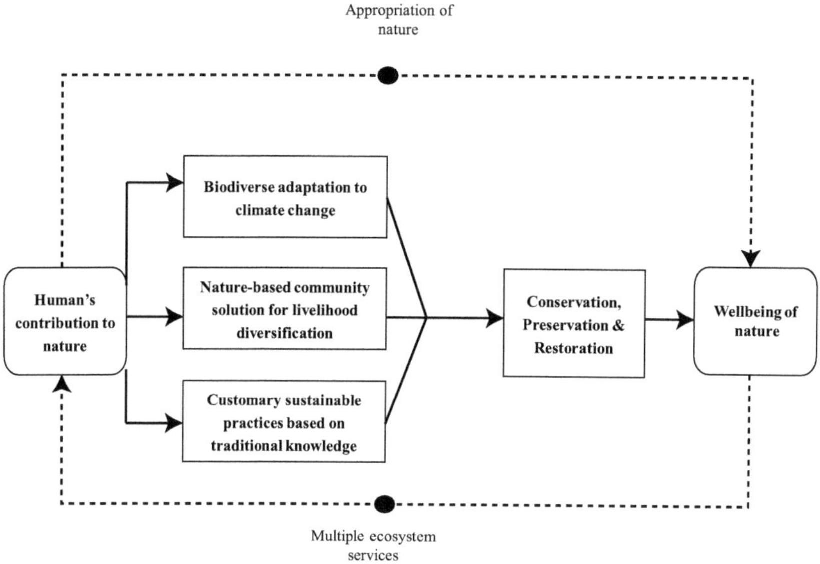

Fig. 12.4 Human's contribution to nature (*Source* Adapted from Titumir & Paran, 2022a)

for a sustainable solution to any shocks in the coastal, marine and forest ecosystem. They have developed course materials on each of the practices for training purposes and also compiled an inventory of traditional knowledge (TK) with the help of *Unnayan Onneshan*. The traditional knowledge is usually acquired through experiences and close contact with the natural environment (Titumir et al., 2020). Alongside, it can be argued that locals know the best strategies for the sustainable usage of the resources. They apply their traditional knowledge, sustainable and customary practices for the conservation of biodiversity and nature. They maintain a few specific rules and practices while harvesting resources that are based on their traditional knowledge. They have some traditional customs and beliefs, which are also consistent with resource conservation. Moreover, they are involved in innovating processes of newer techniques and methods from their own knowledge (Titumir & Afrin, 2017). In the Sundarbans, TK can be categorised based on two different ecosystems, terrestrial ecosystem and coastal and marine ecosystem (see Chapter 4 & 5). TK in the terrestrial ecosystem is followed by woodcutters (*Bawali*), *Golpata* (*Nypa fruticans*) harvesters and honey and wax collectors (*Moual*) while TK in coastal and marine ecosystems is followed by fishermen (*Jele*) and crab collectors.

Innovations in Livelihood Options and Biodiverse Adaptation

In addition to the traditional rules, practices and cultural beliefs that were mentioned above, the TRUs have diversified their livelihood choices by implementing their traditional knowledge and experiences. They regularly attempt to adjust to changing conditions by creating innovative management practices.

Innovative Techniques in Agriculture

The small-scale local farmers have developed some innovative approaches in agriculture that can be adapted to local biophysical conditions while ensuring environmental sustainability. Due to climate change and increase in salinity in soil and water of the coastal region, the farmers grow their rice seedlings in high-rise lands. This is done to lower the risk of saline water contamination and to ensure optimum survival of the seedlings. Then these seedlings are transplanted into the main agricultural land. For example, they cultivate rice plants at 8–12-inch high from the ground level as a response to high salinity content in soil and water. Technically, this saline contaminated rice straw dissolves in a very short time if it is used as the roofing material. Hence, they allow such straws to decompose in the fields which in turn adds organic matter, mainly nitrogen, to the soil and reduces saline intensity. This is favourable for the growth of their next crop. Moreover, the landless people grow vegetables on sheds or roofs, yards or back yards of their houses.

Joint Cultivation of Crab and Duck

As the TRUs often face restrictions while entering the forest for resource harvesting, they use alternative sources of livelihoods. A recent practice is the joint cultivation of crab and duck in one farmland. This practice is by far found to be much profitable for the cultivators. However, intensive research is required for this practice to understand its potential, in terms of both economic and environmental sustainability.

Community-Based Mangrove Agro Aqua Silvi (CMAAS) Culture

The CMAAS culture is referred to the practice of integrated cultivation of some mangrove faunal species—crabs, oyster or fishes (for example, shrimps, *bhetki* [*Latescal carifer*], etc.) and floral species—golpata (*Nypa_fruticans*), *keora* (*Sonerati_aapetala*), *goran* (*Ceriops_decandra*), etc. at the same time on any swampy land of brackish water. Additionally, integrated cultivation of certain mangrove floral species like *golpata* and a few faunal species like *tengra* (*Mystus_tengara*), *baila* (*Awaous_guamensis*), *tilapia* (*Tilapia nilotica*), etc. are practiced in freshwater swampy land. Therefore, CMAAS culture is a substitute for the Commercial Shrimp (CS) culture which has little or no adverse impact on the ecosystem which has been created by the local people (see Chapter 9 more details).

Modified One Health Approach: A Policy Perspective for Post-2020 Biodiversity Framework

'A whole of the society approach' on conservation and sustainable use of biodiversity through SEPLS management in the Sundarbans can ensure and enhance 'one health' that include human, animal and ecosystem health all together. A complete societal approach brings all under one umbrella of practice of sustainable production and consumption as well as equal access and benefit-sharing. It ascertains well-being and good quality of life for human, animal and ecosystem simultaneously. 'One health' is an approach comprising designing and implementing programmes, policies, legislation and research in which several sectors interact and perform together to attain better public health outcomes (WHO, 2017). This approach fundamentally deals with the areas of food safety, control of diseases that spread between animals and human being such as flu, rabies and Rift Valley Fever and fighting antibiotic resistance (WHO, 2017). To assure the effective implementation of 'one health' approach, collaborative efforts and initiatives from multi-sectors are required since there are plethora of dimensions of relationships between animals and human beings in the vast spectrum of ecosystems. According to the Centers for Disease Control and Prevention (CDC), 'one health' approach can fight antibiotic resistance and outbreaks of a zoonotic disease that spread between animal and human and promote food safety and security. As a whole, it works to protect global health security. The uniqueness of this approach is that it identifies the most critical challenges of both current and future contexts and it needs to be implemented by both policy experts and practitioners at field levels. It is largely a participatory proposition.

Thus, promoting 'one health' requires concerted actions. First, ensuring the right-oriented social entitlement for everyone such as universal access to education, health and social security programmes. Second, providing livelihood security that includes food security, job security and social security. Third, promoting human–nature cooperation through green production systems such as clean and green energy, sustainable and ecosystem-based approaches in production and consumption, biodiverse adaptation to climate change, etc. Fourth, promotion of customary sustainable practices and traditional knowledge in SEPLS management (Fig. 12.5).

These activities, however, depend on regional and global partnerships as well as on policy level, production level and ecological conditions. The policy level conditions are indirect drivers of change in the ecosystem that include, as mentioned in Sect. 12.5, institutional and governance system, power and class, property rights and legal arrangements. The production level and ecological conditions are the direct drivers of change in the ecosystem. The production level conditions are carbon emission, pollutions, over-extraction, degradation and exclusion, harvesting and fishing, reforestation and innovative use, etc. while the ecological factors are climate change, weather patterns, natural disasters and hazards. When the drivers of change (production, ecological and political conditions) affect

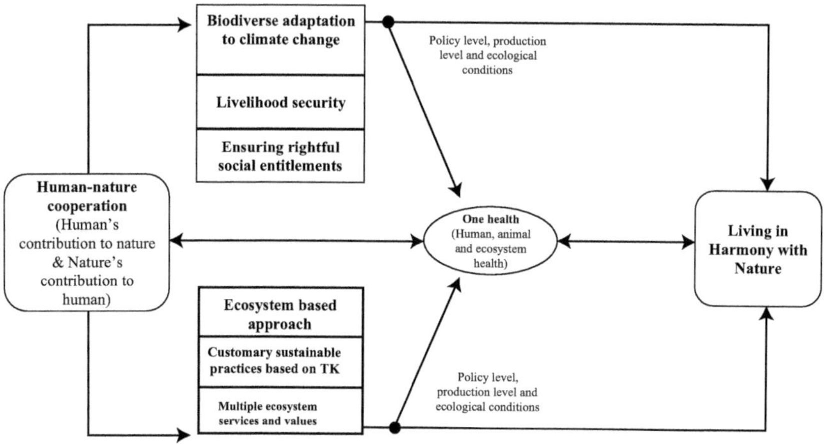

Fig. 12.5 Living in harmony with nature: Modified one health approach for post-2020 period (*Source* Adapted from Titumir & Paran, 2022a)

the ecosystem positively, the multiple services of the ecosystem contribute to human well-being. If these drivers negatively affect the ecosystem, the well-being of both human and nature gets disrupted. It is found that the appropriation of nature (conservation, restoration, sustainable uses access and benefit-sharing) ensures a harmonious relationship between humans and nature which transforms the quality of life of both. Human–nature cooperation counts human's contribution to nature as well as nature's contribution to human beings and contributes to maintaining "one health" through living in harmony with nature (Fig. 12.5).

Conclusions

Exploring the interdependent relationship between humans and nature during the COVID-19 outbreak in the Sundarbans, this chapter reveals that the ecosystem in the Sundarbans is in disruption due to the expropriation of natural resources. As a result, there is massive degradation of biodiversity resources, which is also negatively affecting the life and livelihood of IPLCs. The forest cover is decreasing, therefore negatively affecting the amount and quality of ES. The ongoing pandemic has further exacerbated the socio-economic distress of the forest people. The deprived status of any of the features in the ecosystem—biotic or abiotic—causes a metabolic rift in the ecosystem therefore entrapping the ecosystem health in a vicious cycle. The chapter also finds that the policy response to curb the economic fallout from COVID-19 is inadequate. The absence of universal social security programmes in the midst of livelihood insecurity has heightened the sufferings of the TRUs. It also outlines how nature contributes to human well-being in multiple ways and how human contributes to nature's well-being significantly.

Analysing the community understanding on one health approach, the study demonstrates that human and nature are dependent on each other forming a human–nature sociality in the ecosystem through which they coexist. Therefore, the appropriation of nature (conservation, restoration, sustainable uses, access and benefit-sharing) instead of expropriation (anthropogenic pressures) is the yardstick to ensure virtuous cycle (good quality of life) in the ecosystem and harmonious relations between human and nature. In this regard, this chapter presents a modified 'one health' framework for a post-2020 period which can promote human transition into living in harmony with nature. The framework calls for ensuring right-oriented universal social entitlement, provision of livelihood security and promotion of human–nature cooperation through green production, ecosystem-based approach and biodiverse adaptation to climate change, underwritten by customary sustainable practices and traditional knowledge in SEPLS management.

Acknowledgement The chapter draws on 'Human-Nature Cooperation for Well-being: Community Understanding on One Health Approach in the COVID-19 Era in the Sundarbans', published by Springer, Singapore in M. Nishi, S. M. Subramanian, H. Gupta (Eds.), 2022, *Biodiversity-Health-Sustainability Nexus in Socio-Ecological Production Landscapes and Seascapes (SEPLS)* as a part of the Satoyama Initiative Thematic Review book series of United Nations University Institute for the Advanced Study of Sustainability and 'Ecosystem services and well-being in the Sundarbans of Bangladesh: A multiple evidence base trajectory' to be published by Springer Nature, Singapore in R. Dasgupta, S. Hashimoto & O. Saito (Eds.), 2022, *Assessing, mapping and modelling of mangrove ecosystem services*.

References

Islam, M. S. N., & Gnauck, A. (2009). Threats to the Sundarbans mangrove wetland ecosystems from transboundary water allocation in the Ganges basin:

A preliminary problem analysis. *International Journal of Ecological Economics and Statistics, 13*(9), 64–78.
IPBES. (2019). *Summary for policymakers of the global assessment report on biodiversity and ecosystem services of the Intergovernmental Science-Policy Platform on Biodiversity and Ecosystem Services* (Eds., S, Díaz, J, Settele, ES, Brondízio, HT, Ngo, M, Guèze, J, Agard, A, Arneth, P, Balvanera, KA, Brauman, SHM, Butchart, KMA, Chan, LA, Garibaldi, K, Ichii, J, Liu, SM, Subramanian, G,F, Midgley, P, Miloslavich, Z, Molnár, D, Obura, A, Pfaff, S, Polasky, A, Purvis, J, Razzaque, B, Reyers, R, Chowdhury, YJ, Shin, IJ, Visseren-Hamakers, KJ, Willis & CN, Zayas, IPBES secretariat, Bonn, Germany.
Titumir, R. A. M., & Paran, M. S. (2022a). Human-nature cooperation for well-being: Community understanding on one health approach in the COVID-19 era in the Sundarbans. In M. Nishi, S. M. Subramanian, & H. Gupta (Eds.), *Biodiversity-health-sustainability nexus in socio-ecological production landscapes and seascapes (SEPLS).* Springer Nature.
Titumir, R. A. M., & Paran, M. S. (2022b). Ecosystem services and well-being in the Sundarbans of Bangladesh: A multiple evidence base trajectory. In R. Dasgupta, S. Hashimoto & O. Saito (Eds.), *Assessing, mapping and modelling of mangrove ecosystem services.* Springer Sustainability Science Book Series (Forthcoming).
Titumir, R. A. M., & Paran, M. S. (2022c). Human-nature relationship for post 2020 biodiversity framework: An exploration into traditional knowledge, customary Sustainable use and community-based innovation in the Sundarbans. In R. A. M. Titumir (Ed.), *Fifty years of nation building: Political economy of Bangladesh's development.* Centennial Book Series Project of University of Dhaka. Department of Development Studies, University of Dhaka (Forthcoming).
Titumir, R. A. M., Afrin, T., & Islam, M. S. (2022). *Natural resource degradation and human-nature wellbeing: Cases of biodiversity resources, water resources, and climate change.* Springer (Forthcoming).
Titumir, R. A. M., Afrin, T. & Islam, M. S. (2020). Traditional knowledge, institutions and human sociality in sustainable use and conservation of biodiversity of the Sundarbans of Bangladesh. In O. Saito, S.M. Subramanian, H. Hashimoto & K. Takeuchi (Eds.), *Managing socio-ecological production landscapes and seascapes for Sustainable Communities in Asia: Mapping and Navigating Stakholders, Policy and Action* (pp. 67–92). Science for Sustainable Societies. https://doi.org/10.1007/978-981-15-1133-2_5
Titumir, R. A. M., Paran, M. S. & Pasha, M. W. (2019), The Sundarbans is our mind: an exploration into multiple values of nature in conversation with traditional resource users', in *Understanding the multiple values associated with sustainable use in socio-ecological production landscapes and seascapes (SEPLS)*, Satoyama Initiative Thematic Review, vol. 5, eds. UNU-IAS & IGES, United

Nations University Institute for the Advanced Study of Sustainability, Tokyo, Japan, pp. 97–117.

Titumir, R. A. M., & Afrin, T. (2017). Complementarities of human-nature well-beings: A case illustrated through traditional forest resource users of Sundarbans in Bangladesh. *Satoyama Initiative Thematic Review (SITR)—Volume 3*, United Nations University Institute for the Advanced Study of Sustainability (UNU-IAS): Tokyo, Japan.

Unnayan Onneshan. (2020a). *Biodiversity, climate change and traditional resource users in the Sundarbans: An exploration through public participation geographic information system (PPGIS)* (Unpublished Research Report).

Unnayan Onneshan. (2020b). Whither bending for life and livelihood: A rapid assessment of national budget 2020b-21. *Bangladesh Economic Update, 11* (9), Unnayan Onneshan, Dhaka, Bangladesh.

Unnayan Onneshan. (2020c). *Promoting diverse cultural values of biodiversity ecosystem services.* Presentation to FPP virtual partners' meeting November 11. Unnayan Onneshan, Dhaka, Bangladesh.

World Health Organization (2017). *One health.* Available at https://www.who.int/news-room/questions-and-answers/item/one-health

CHAPTER 13

Conclusions: A Post-2020 Biodiversity Takeaway

Rashed Al Mahmud Titumir

The Sundarbans which is the largest single tract mangrove forest in the world and a World Heritage Site located along the Bangladeshi coast is undergoing a fast-paced transformation. In the last two decades from 2000–2020, the mangrove forest that serves as a wall against many natural phenomenon has degraded losing both its density and size. Many areas in the forest are turning into fallow lands. The area comprising the dense forest area has decreased by almost half. In the first decade from 2000 to 2010, deforestation on a large scale reduced the vegetation density and reduced forest cover. However, the growth rate of fallow land was relatively low at that time. But in the following decade from 2010–2020, the area of fallow land has almost doubled. Although vegetation can be seen around the forest, there is widespread fallow areas inside the forest.

R. A. M. Titumir (✉)
Department of Development Studies, University of Dhaka, Dhaka, Bangladesh
e-mail: rt@du.ac.bd; rtitumir@unnayan.org

Unnayan Onneshan, Dhaka, Bangladesh

R. A. M. Titumir (ed.), *Sundarbans and its Ecosystem Services*, Sustainable Development Goals Series,
https://doi.org/10.1007/978-981-19-3000-3_13

Currently, the area of forest has been reduced to almost half of what it was in 1776. Both aquatic, floral and faunal species have significantly declined in terms of quality and quantity. The value of waterbodies based on the availability of aquatic resources like large fish and crabs, have been gradually decreasing in the last two decades, which implies that resource is being exploited. The profit-making impulse of people has led to extraction of the resources in an unsustainable manner. These negative changes are impacting the quality of the forest, its ability to deal with natural disasters, climate change and the ability to reduce air pollution. In addition to all these, the natural beauty of the forest is declining too.

This book found that ecosystem in the Sundarbans is getting disrupted due to expropriation of natural resources. As a result, there is massive degradation of biodiversity resources, which is negatively affecting the life and livelihood of Indigenous People and Local Communities (IPLCs) (see Titumir et al., 2022; Titumir & Paran, 2022a, 2022b, 2022c for more). The COVID-19 pandemic has further exacerbated the socio-economic distress of the forest people (see Titumir & Paran, 2022a for more). The existing regime responsible for policy making and implementing forest management system has been found to be irresponsive and inadequate to maintain the ecosystem balance in the Sundarbans and curb the hardship of the IPLCs. The absence of universal social security programmes coupled with livelihood insecurity has heightened the sufferings of people. The book outlines that nature contributes to human well-being in multiple ways and vice-versa. The research found that customary sustainable practices of Traditional Resource Users (TRUs) based on their traditional knowledge can promote adaptation and make the forest resilience towards climate change and other natural shocks. The book, therefore, calls for acknowledging those traditional knowledge and practices in formal forest management system and strengthening of co-management in the forest involving IPLCs including TRUs. It demonstrates that human and nature are dependent on each other and forms a human–nature sociality in the ecosystem through which they coexist. Therefore, it is necessary to promote appropriation of natural resources instead of expropriation. In this regard, it presents a modified one health framework for the post-2020 period, which can promote transition among humans and make them live in harmony with nature.

The forest is losing its biodiversity. As the forest cover is shrinking, the number of plant species is also declining. The Sundari (Heritiera fomes) is the fastest declining tree species. The existence of various animals in the forest is also under threat due to declining plant cover. The average height of trees in the forest is decreasing (see Titumir & Paran, 2022a). As a result, the habitat of many species of birds, monkeys and tree-dwelling species is under threat. Marine and aquatic resource of the Sundarbans are also in a state of eternal decline as resource extraction has gone up from what it used to be earlier. Most of the endangered species are finfish and shellfish. The Sundarbans is home to more than 31 species of globally endangered wildlife . According to the IUCN, more than 40 species of amphibians, reptiles, birds and mammals are considered endangered or critically endangered. The globally endangered *Batagur baska* tortoise is found in the Sundarbans but now it is critically endangered. Gangetic dolphins, Irrawaddy dolphins and saltwater crocodile species are also endangered. Researches published in various journals contain different numbers of wild animals in the Sundarbans. But no other animals are regularly surveyed in the Sundarbans except tigers and dolphins. As a result, it has become difficult to ascertain the exact number of these animals left in the forest. Regularly updated statistics on the number of biodiversity resources of the Sundarbans should be published following the accepted scientific method. In this regard, the government can involve IUCN and its commission members.

Biodiversity in the Sundarbans is being lost due to uncontrolled extraction of forest resources without thinking about forest biodiversity, conservation and reproduction (see Titumir et al., 2019, 2020 for more). Due to the high demand and price of forest resources in the market, a class of people are engaging in unsustainable means of resource extraction. For example, according to traditional resource users in the Sundarbans, the number of valuable trees, honey and large fish is declining due to illegal extraction without considering conservation and reproduction. The woodcutter community of the Sundarbans known as bawalis said that the extraction of valuable trees from different parts of the Sundarbans has been increasing in the last two decades from 2000–2020, leading to a gradual decline in the number of valuable trees. Fishermen said that the number of large fish in different regions has decreased in two decades (2000–2020). Mouals, the honey and wax collectors, are also saying the same that the amount of honey and wax available in different regions has also decreased in two decades.

On the contrary, traditional rules and practices upheld by the TRUs, while collecting resources from the forest, instinctively leads to the conservation of the forests' ecosystem and biodiversity and helps in the sustainable and proper use of scarce resources (see Titumir & Afrin, 2017; Titumir et al., 2020). The cultural beliefs of the TRUs act as the founding stone for the forms of their practices. TRUs utilise their age-old conservational rules and practices whilst gathering resources from that belt. In this book, the rules and practices followed by the TRUs have been specified with the notion of corroborative evidence. It becomes clear that, the way they collect honey, crab, shell, fish, golpata and other resources are favourable for both nature and human.

In extracting forest resources, they follow certain methods and rules. For example, mouals (honey and wax collectors) in the Sundarbans collect honey in April, May and June. They cut a certain part (about two-thirds) of the beehive. The rest is left for the next breeding season. They collect honey by hand without using metal tools, so that the young bees do not die. In this process they make smoke using dried leaves found in the forest and never put fire in the beehives. As a result, the bees are less likely to die and can reuse the hive.

Similarly, golpata collectors follow certain rules. They do not collect golpata from the same region more than once a year and collection is prohibited during the breeding period from June to September. They cut the leaves that are about nine feet long and cut them in such a way that the central leaf of each tree and its side leaves are not damaged. While cutting the leaves, it is ensured that no damage is done to the flowers, fruits and the young trees. In this process, the preservation and healthy reproduction of golpata is ensured. But outsiders do not know these traditional techniques. As a result, they collect forest resources in an exploitative manner.

Moreover, the Community Managed Agro Aqua Silvi Culture (CMAAS) culture has been introduced following traditional knowledge and customary sustainable practices by the TRUs, aiming to conserve the resources undermining shrimp culture that is proving harmful for other floral and faunal species. The CMAAS culture practiced within the forest people communities as a part of their conservation criteria is economically profitable, environmentally friendly and socially acceptable, compared to other methods of cultivation i.e., commercial shrimp cultivation.

The Sundarbans provides innumerable services. It also has a role to play in highlighting the natural beauty, dignity and heritage of Bangladesh on the world stage. It saves the country against climate change by acting as a shield against many climatic hazards. This role of the mangrove forest acting as a climatic shield necessitates conserving it with greater care in the face of ever-deteriorating climatic conditions and global phenomenon of climate change. The forest provides multiple benefits and services that are crucial to cope with the vulnerabilities that result from climate change. The Sundarbans' ecosystem services and value contribute to human well-being. Different types of ecosystem services thrive if nature's well-being is maintained.

True value of the forest cannot be proven through mere market pricing, because this market-centric approach will fail to determine the forest's worth. Additionally, natural resources should be evaluated through the eyes of traditional resource users . The TRUs consider the Sundarbans to be their mind, which means they equate their life and well-being with the life and well-being of the forest as well as their thinking process and way of life around the life and spirit of the forest (see Titumir et al., 2019; Titumir & Paran, 2022a). TRUs' conceptualisation of value challenges the orthodoxy, particularly of the neo-classical notion of marketisation of natural resources. This is because the forest includes multiple value and ecosystem services which are beyond monetary entitlements. The value is important for both human and nature's existence. On top of this, the book proposes that formal conservation practitioners should follow the TRUs' way of valuation of nature. This mangrove forest is not a mere foliage occupying a large portion of inland and marine areas, but it is a natural shield or protector which has been protecting the resources of the ecosystem over the years. It works as a carbon reservoir. Moreover, it fights impulsively against natural disasters, salinity, erosion, sea-level rise and other calamities. Hence, it is vital to save this forest for the sake of nature, human and other resources.

Numerous forms of customary knowledge have been applied by local people to protect the forest. They have formed cooperatives to claim their rights where conservation of the natural resources of the forest occupies a big portion. The problem is, as per their accusation, the forest department, local politicians and police who are not in favour of their rigorous efforts. Institutional fragility and corruption of the officials are giving way to expropriation of resources. Masses are getting alienated from the natural resources. As a consequence, commodity fetishism is on the rise, which reflects the motive of the clientelist groups.

The case of three forest people cooperatives, dealing with various socio-political hurdles towards achieving and claiming their rights demonstrates that traditional knowledge system can significantly promote sustainable management of natural resources, both within the protected areas system and other effective area-based conservation measures, if they are acknowledged and supported by government and non-government agencies. If IPLCs are empowered, they can significantly conserve and restore the biodiversity of the forest using their Indigenous and Local Knowledge (ILK) and customary sustainable practices. However, there is lack of government's recognition and adoption of traditional knowledge in formal rules and laws of the forest management.

This book identified several customary sustainable practices by TRUs based on traditional knowledge, to be promoted as a long-term community-based adaptation and resilience measure to climate change and other shocks i.e., COVID-19. It is, therefore necessary to acknowledge those traditional knowledge and practices, which can initiate innovative community-based solutions against climate change and enhance the resilience capacity of the community. On top of that, a whole of the social approach is needed to bring everyone under the practices of sustainable production and consumption as well as equal access and benefit-sharing, which ascertains well-being and good quality of life for human, animal and environment simultaneously as a part of 'one health approach'. However, upscaling and promoting such approaches as a long-term adaptation measure is a big challenge due to shortage of funding and lack of political will. If the TRUs get the support of the government and non-government agencies, they will be able to contribute a lot towards the forest assisting in maintaining its valuable resources and biodiversity. However, the presence of overlapping property rights regime—often marginalising the TRUs leads to conflict between management of resources and its degradation.

The ecosystem and biodiversity of the Sundarbans is under extreme threat mainly due to anthropogenic pressures, climate change and natural disasters. First, politically and economically influential groups are illegally occupying forests. Human settlements are gradually increasing around the forest. Forests are being deforested due to illegal logging even though this is a declared natural sanctuary. Second, commercial activities are increasing day by day around the forest even in the ecologically critical areas. Third, biodiversity is being lost due to over-extraction of forest resources. Again, many species are getting lost due to exploitation of resources in harmful ways. Fish are collected using harmful nets and poisoning. Fourth, even when it comes to co-management, the existing forest management system is gradually marginalising the rights and traditional knowledge of IPLCs or TRUs. For example, each time they have to take a clearance from the forest department with a certain amount of money to extract resources. There are allegations of irregularities and corruption in this process. As a result, in many cases, TRUs are engaged in amassing resources beyond the amount required for survival. Fifth, biodiversity and ecosystems are being harmed by frequent natural disasters. Sixth, due to climate change, various organic and inorganic components of forests (such as salinity, rainfall, pH, mineral content, etc.) are changing, posing a negative impact on the ecosystem.

This largest single tract and composite mangrove forest is at risk due to the relentless and illegal extraction of resources by a group of influential people, who have created a clientelist network among themselves to gain high profit. Before the rise of expropriation, the natural harmony was sound and resilient through the human sociality and through an interdependent relationship between human and nature. The traditional knowledge used by the forest people is beneficial for both humans and nature. These techniques of resource appropriation help to conserve the resource of both terrestrial and marine regions. The TRUs continue their efforts to protect the forest, which they consider to be their mother using inherent sustainable knowledge. However, their voices are not heard in the formal co-management process. Thus, the things in the forest are falling apart, and the concern of destruction is increasing by the day.

Institutional fragility is a big crisis against the backdrop of several illegal expropriations. The government should make and enact rules properly acknowledging rights of the TRUs . Forest people should be included in the co-management process. Their methods of conserving the forest and valuation of resources should be particularly considered in the conservation module which is undertaken by the formal authority. At present, TRUs have cooperatives which are working as a feasible platform to raise their voice. They are now going ahead proclaiming their needs and demands. TRUs are learning to negotiate, make pressure and create movement to claim their rights, which also includes the conservation of the forest.

The increase involvement of local people in conservation and restoration activities will lead them to act as the watchdog to stop the hunting and trafficking of wild animals and plants. Moreover, they will refrain from doing things that are harmful to wildlife. For this, it is necessary to make the people aware of wildlife conservation, breeding, etc. The conservation, restoration and preservation of the Sundarbans will be accelerated if traditional knowledge and customary sustainable practices of IPLCs or TRUs can be enhanced through recognising them in formal rules and practices. If there are opportunities for participation of IPLCs, people from different classes, professions and civil society members, skills and transparency in forest management will increase. However, forest laws need to be modernised and enforced to conserve forests, nature and biodiversity. The government has sought public opinion on various aspects of the draft Forest Act 2019. Some views and proposals have already been put forward by different stakeholders. Adherence to various international laws is also important in modernising forest laws. Strict action needs to be taken against deforestation, encroachment and any activities that are detrimental to forests and its biodiversity. In this case, proper implementation of the law is important.

The COVID-19 pandemic teaches the importance of nurturing the interrelations between nature and human being. It is very important to protect the biodiversity of the Sundarbans for the sake of human's own existence. Biodiversity can be saved from extinction if the interdependence and cooperation between human and nature are restored. Through this process, the balance of the environment will be maintained and the well-being of both people and the environment will be ensured. Wildlife conservation requires maintaining the balance of the ecosystem. The natural environment contains inorganic matter, living matter and microorganisms. Every element is important for a balanced ecosystem. When any one element of the ecosystem is damaged due to physical, chemical and biological reasons, the balance is lost. As a result, the food chain of producers, consumers and microorganisms is broken. That is why different species of plants and animals are lost. Hence, there is no substitute for keeping the cyclical chain of ecosystems by conserving every element. Just as man is responsible for the destruction of nature, so nature will regain its life through the efforts of man. A sense of mutual cooperation and interdependence will be awakened through the re-establishment of the interrelationship between human and nature. This valuable forest needs to be made free from power plants and industries. The government needs to take prompt initiative, with a provision of strict enforcement of rules and laws to save the forest. It should be realised that well-being is just as necessary for nature, as it is for humans. This intertwined relationship of well-being, will help to ensure sustainable natural ambience and harmony between the people and nature.

In order to accelerate the conservation, restoration and preservation measures, the following issues need to be kept in mind. First, small scale tree planting and restoration projects can have a big impact on the environment. There are even examples of changes in development design to save a tree in different countries of the world. Second, forest life must be saved to improve human mental and physical well-being, reduce air pollution, cool the environment and increase the quality of other services from the forest. Third, it is necessary to make decent and environment friendly livelihood through forest conservation. A sustainable and environmentally friendly production system can be ensured if forests are conserved and restored through proper management. As a result, new employment will be created. Fourth, the conservation and restoration of forests requires the involvement and empowerment of local people. Co-management activities need to be strengthened with the participation of IPLCs. Then the local people will come forward to protect the forest and forest biodiversity. Moreover, they will refrain from doing things that are harmful to the forest. Fifth, more joint management projects like IUCN's Tanguar Haor, Cox's Bazar and Nijhum Island can be undertaken to protect the biodiversity of different ecosystems through the combined efforts of government, private and local people. Through joint initiatives across the country, the same measures can be taken in all ecosystems to reduce the risk to some extent. Sixth, the protection of forests and forest biodiversity is essential for peoples' own survival. People will live only if the forest survives. There is no substitute for forest conservation and restoration for a balanced environment and healthy and beautiful life.

The interconnectedness of human and nature, along with the well-being resulting from the benefits received from ES subjected to production, ecological and political conditions. The policy level conditions are institutional and governance system, power and class, property rights and legal arrangements. If these factors function inclusively, they positively affect the ecosystem's health. In contrast, if policy level conditions take extractive nature, it causes disruption in the ecosystem and thus ill-being of human, animal and ecosystem. The production level conditions include the level of carbon emission and pollutions, the rate of extraction, degradation and exclusion,the nature of harvesting and fishing, the rate of reforestation and innovative use etc. If production and consumption move forward in sustainable and green processes, the quality of the ecosystem remains healthy. Otherwise, anthropogenic pressures lead to massive depletion of the nature and thus ill-being of all biotic and abiotic features of the ecosystem. On the other hand, the ecological factors are climate change, weather patterns, natural disasters and hazards, which also determine the provision of ES and the level of well-being of ecosystem. When the drivers of change (production, ecological and political conditions) affect the ecosystem positively, the services of the ecosystem derive from nature's well-being to contribute to human well-being in many forms. If the drivers negatively affect the nature, then well-being of both nature and human being gets disrupted. The appropriation of nature (conservation, restoration, sustainable uses access and benefit-sharing) ensures a harmonious relationship between human and nature, which transforms the quality of life of both. Every component of the ecosystem must be allowed to live in its own sphere. Every part of the ecosystem is crucial for each other's survival, and hence they need the freedom to thrive. Promoting social freedom—rights of all components of the ecosystem to live fully without causing harm to other—will help maintain a balance in the ecosystem, contributing to well-being of all (see Brand et al., 2021 for more).

In contrast to resource expropriation, sustainable use of natural resources is healthy. It is also crucial to strengthen the ILK and customary practices of the TRUs, who consider the forest as their mind. Their understanding suggests that socially constructed norms and values—the informal institutions can solve the management problem of natural resources. Since TRUs consider natural resources of the forest as their intimate part of life, they seldom cause harm to them. Hence, incorporating their norms and values in the formal conservation process is very important to save natural resources.

The resilience factors and contributions of the Traditional Resource Users (TRUs) towards the SDGs and the Aichi Biodiversity Targets reveal that if the TRUs get the support of the government and non-government agencies, they will be able to contribute a lot towards the forest in maintaining its valuable resources and biodiversity. The degradation of natural resources was not only about the non-existence of market, but also about the distorted power sharing by the stakeholders over the management of resources. The presence of overlapping property rights regime leading to conflicts between management of resources and its degradation. The conservation framework based on human sociality led to a better outcome. This is because it led to the sustainable utilisation and production of biodiversity resources, where the amount that was yield was greater than the amount harvested. This was in contrast to the outcomes under the patron-client relationship which resulted in an unsustainable biodiversity resource extraction that led to a greater amount of harvest compared to a relatively lower amount of yield. To sum it up, this acted as an incentive to conserve the knowledge and ensure sustainable value chain which will boost up a good practise and help maintain resources.

The massive degradation of biodiversity resources also negatively affects the life and livelihood of IPLCs. The forest coverage is decreasing, therefore negatively affects the amount and quality of ES. The ongoing pandemic has heightened the socio-economic distress of the forest people. The ill-being status of any of the features in the ecosystem—biotic or abiotic—causes a metabolic rift in the ecosystem and therefore entraps the ecosystem health in a vicious cycle. The policy response to curb the economic fallout from COVID-19, without having a universal social security programmes in the midst of livelihood insecurity is inadequate. Therefore, 'a whole of the society approach' on conservation and sustainable use of biodiversity through SEPLS management in the Sundarbans can ensure and enhance 'one health' that include human, animal and ecosystem health all together.

It is to mention that, around the post-2020 Global Biodiversity Framework (GBF) and the Paris Agreement on climate change, the prominent issue in the discourse on the biodiversity and climate change crises is the 'Nature-based Solutions (NbS)'. Governments, international organisations and other non-governmental organisations are supporting NbS and using it interchangeably with other terms like 'natural solutions', 'natural climate solutions' and 'ecosystem-based approaches' to describe an array of positive actions and approaches such as agro-ecology and ecosystem restoration. Besides, there is also growing participation of IPLCs in the CBD process with a view to bringing issues and demands of IPLCs on fore and getting the recognition of their ILK and entitlements as noted in article 8 (j) of CBD. The recognition of the knowledge, values, roles and contributions of IPLCs to biodiversity conservation and sustainable use has been achieved during the past decade, however IPLCs and indigenous and local knowledge remain marginalised and largely invisible, especially in policy and decision-making at the national level. This is a major missed opportunity in seeking solutions to the worsening biodiversity-climate-development interlinkages crisis. The call for 'transformational change' by UN processes such as IPBES, UNFCCC and CBD call cannot be achieved unless IPLC collective actions are fully recognised and supported. Promoting human-rights-based approaches and NbS will continue to be a leading principle and approach in upcoming days. There will be a shift in focus from the inputs into the post-2020

Global Biodiversity Framework to the follow-up and implementation of the framework at national and local levels.

Against this backdrop, a work package incorporating a holistic approach, promoting NbS and IPLCs active participation and leadership can be promoted to introduce a whole of the social approach as well conservation, restoration and preservation in the Sundarbans in post-2030 period. These activities need to aim at improving communities' resilience and response to emerging risks and changing circumstances, including the COVID pandemic situation as well as addressing the issues in global and regional policy forums, such as the CBD and IPBES, by highlighting the underlying inequalities that generate disproportionate impacts on vulnerable groups such as IPLCs.

A whole of the society approach brings all under the umbrella practices of sustainable production and consumption as well as equal access and benefit-sharing ascertaining well-being and good quality of life for human, animal and ecosystem simultaneously. Promoting 'one health' requires concerted actions for ensuring universal access to education, health and social security programmes, providing food security, job security and social security, promoting human-nature cooperation, ecosystem-based approach in production and consumption, and customary sustainable practices and traditional knowledge in SEPLS management.

References

Brand, U., et al. (2021). From planetary to societal boundaries: an argument for collectively defined self-limitation. *Sustainability: Science, Practice and Policy*, *17*(1), 265–292. https://doi.org/10.1080/15487733.2021.1940754

Titumir, R. A. M., Afrin, T., & Islam, M. S. (2020). Traditional knowledge, institutions and human sociality in sustainable use and conservation of biodiversity of the Sundarbans of Bangladesh. In O. Saito, S. M. Subramanian, H. Hashimoto, & K. Takeuchi (Eds.), *Managing socio-ecological production landscapes and seascapes for sustainable communities in Asia: Mapping and navigating stakholders, policy and action* (pp. 67–92). Science for Sustainable Societies. https://doi.org/10.1007/978-981-15-1133-2_5

Titumir, R. A. M., & Afrin, T. (2017). Complementarities of human-nature well-beings: A case illustrated through traditional forest resource users of Sundarbans in Bangladesh. In *Satoyama Initiative Thematic Review (SITR)—Volume 3*. United Nations University Institute for the Advanced Study of Sustainability (UNU-IAS).

Titumir, R. A. M., Afrin, T., & Islam, M. S. (2022). *Natural resource degradation and human-nature wellbeing: Cases of biodiversity resources, water resources, and climate change*. Springer (Forthcoming).

Titumir, R. A. M., & Paran, M. S. (2022a). Human-nature cooperation for well-being: Community understanding on one health approach in the COVID-19 era in the Sundarbans. In M. Nishi, S. M. Subramanian, H. Gupta (Eds.), *Biodiversity-health-sustainability nexus in socio-ecological production landscapes and seascapes (SEPLS)*. Springer Nature, Singapore.

Titumir, R. A. M., & Paran, M. S. (2022b). Ecosystem services and well-being in the Sundarbans of Bangladesh: A multiple evidence base trajectory. In R. Dasgupta, S. Hashimoto, & O. Saito (Eds.), *Assessing, mapping and modelling of mangrove ecosystem services*. Springer Sustainability Science Book Series.

Titumir, R. A. M., & Paran, M. S. (2022c). Human-nature relationship for post 2020 biodiversity framework: An exploration into traditional knowledge, customary sustainable use and community-based innovation in the Sundarbans. In R. A. M. Titumir (Ed.), *Fifty years of nation building: Political economy of Bangladesh's development*. Centennial Book Series Project of University of Dhaka. Department of Development Studies, University of Dhaka.

Titumir, R. A. M., Paran, M. S., & Pasha, M. W. (2019). The Sundarbans is our mind: An exploration into multiple values of nature in conversation with traditional resource users. In UNU-IAS & IGES (Eds.), *Understanding the multiple values associated with sustainable use in socio-ecological production landscapes and seascapes (SEPLS)* (Satoyama Initiative Thematic Review, Vol. 5, pp. 97–117). United Nations University Institute for the Advanced Study of Sustainability.

Index

R. A. M. Titumir (ed.), *Sundarbans and its Ecosystem Services*, Sustainable Development Goals Series,
https://doi.org/10.1007/978-981-19-3000-3

H

I

J

L

M

GPSR Compliance

The European Union's (EU) General Product Safety Regulation (GPSR) is a set of rules that requires consumer products to be safe and our obligations to ensure this.

If you have any concerns about our products, you can contact us on ProductSafety@springernature.com

In case Publisher is established outside the EU, the EU authorized representative is:

Springer Nature Customer Service Center GmbH
Europaplatz 3
69115 Heidelberg, Germany

Batch number: 09576550

Printed by Printforce, the Netherlands